SPON'S CONSTRUCTION RESOURCE HANDBOOK

Also available from E & FN Spon

Spon's Architects' and Builders Price Book
Davis, Langdon and Everest

Spon's Mechanical and Electrical Services Price Book
Davis, Langdon and Everest

Spon's Civil Engineering and Highway Works Price Book
Davis, Langdon and Everest

Spon's Landscape and External Works Price Book
Davis, Langdon and Everest

Spon's House Improvements Price Book
B. Spain

Spon's Building Costs Guide for Educational Premises
Second Edition
D. Barnsley

Spon's Landscape Handbook
Fourth Edition
Derek Lovejoy Partnership

HAPM Component Life Manual
HAPM Publications Ltd.

Green Building Handbook
T. Woolley, S. Kimmins, R. Harrison and P. Harrison

Procurement Systems: A Guide to Best Practice in Construction
S. Rowlinson and P. McDermott

Site Management of Building Services Contractors
J. Wild

Understanding JCT Standard Building Contracts
Fifth Edition
D. Chappell

Understanding the Building Regulations
S. Polly

Building Regulations Explained
Fifth Edition
J. Stephenson

To order or obtain further information on any of the above or receive a full catalogue please contact:
The Marketing Department, E & FN Spon, 11 New Fetter Lane, London EC4P 4EE.
Tel: 0171 842 2400; Fax: 0171 842 2303

SPON'S CONSTRUCTION
RESOURCE HANDBOOK

BRYAN SPAIN

E & FN Spon
An imprint of Routledge

London and New York

This edition published 1998
by E & FN Spon, an imprint of Routledge
11 New Fetter Lane, London EC4P 4EE

Simultaneously published in the USA and Canada
by Routledge
29 West 35th Street, New York, NY 10001

© 1998 E & FN Spon

Printed and bound in Great Britain by
TJ International Ltd, Padstow, Cornwall

Publisher's Note
This book has been prepared from camera-ready provided by the author.

British Library Cataloguing in Publication Data
A catalogue record for this book is available
from the British Library

ISBN 0 419 23680 5

Contents

Preface   ix

Introduction   xi

PART ONE   BUILDING   1

1   Excavation and filling   3

2   In situ and precast concrete   21

3   Brickwork and blockwork   39

4   Stonework   61

5   Asphalt work   67

6   Roofing and cladding   71

7   Carpentry and joinery   97

8   Structural steelwork   125

9   Metalwork   135

10   Floor finishes   139

11   Wall and ceiling finishes, partitions   153

12   Plumbing   175

13   Glazing   197

14   Wallpapering   211

15   Painting   217

16   External works   233

17	Drainage	243
18	Elemental percentage breakdowns	257

PART TWO   CIVIL ENGINEERING   265

19	Demolition	267
20	Site clearance	269
21	Excavation and filling	271
22	Geotextiles	287
23	Concrete work	289
24	Drainage	309
25	Roads	347
26	Brickwork, blockwork and masonry	359
27	Timber	379
28	Painting	393
29	Waterproofing	405
30	Fencing	409
31	Gabions	419

PART THREE   LANDSCAPING   421

32	Seeding and turfing	423
33	Trees	429
34	Land drainage	435
35	Water supply and ponds	441

PART FOUR   MECHANICAL WORK   445

36   Piped supply systems   447

37   Heating and cooling systems   467

38   Ventilation systems   507

PART FIVE   ELECTRICAL WORK   513

39   General electrical information   515

40   Conduits, cable trunking and trays   519

41   Transformers, switchgear and
distribution boards   529

42   Luminaires and lamps   539

43   Electrical accessories   547

44   Fire detection and alarms   551

45   Earthing and bondings   553

46   Cables and wiring   557

PART SIX   GENERAL INFORMATION   575

47   Measurement data   577

48   Useful addresses   583

Index   591

Preface

The purpose of this book is to provide data for use in the preparation of analytical rates in the construction industry. Although there is a perception that there is a definitive rate for, say, a square metre of walling constructed of common bricks in cement mortar 112mm thick, this is not the case. Men work at varying rates, materials are bought at varying discounts and contractors seek to recover varying levels of overheads and profit.

Nevertheless, it is vitally important both for contractors and clients, that estimates are based upon the most accurate information available. This book's aim is to provide data for all aspects of the industry including building, civil engineering, landscaping, mechanical and electrical work. In addition to the provision of labour and plant constants or norms, further information is set out to assist in the ordering of materials.

This book is presented generally in the style and format of standard methods of measurements but there are many cases where I thought a departure would assist in presenting the data in a more helpful way.

I gratefully acknowledge the assistance I have received from many contractors and material manufacturers in the preparation of this book. In particular, I would like to thank the Rosebery Group, Trada Technology, Rotary North West and Richard Kimpton of Kimpton Queensway for their help.

I would welcome constructive criticism together with suggestions for improving the scope of the contents.

Although every care has been taken in the research and preparation of the data, neither the publishers nor I can accept any liability of any kind resulting from the use of the data provided.

Bryan J D Spain
E & FN Spon
11 New Fetter Lane
London EC4P 4EE

Introduction

The vigorous pursuit by private and public clients for value for money in the construction industry has never been more pronounced. This has led to the need for a reassessment of analytical rate build-ups to ensure that the core data being used - labour and plant constants - are accurate.

These constants must be consistent and their validity proven by use and observation. They should only require minor adjustments in non-standard circumstances, such as small packages of work, difficult working conditions or where an unusually high standard of finish is required.

The contents of this book are intended to provide such a base. The book is presented under the following main headings:

- building;

- civil engineering;

- landscaping;

- mechanical;

- electrical.

Where appropriate, constants are also given for plant, and other helpful information is included to assist in the ordering of materials. The data provided are based upon the premise that the item descriptions refer to contracts large enough to warrant the establishment of site hutting and non-working supervisors.

Contracts of this size will have separate preliminary items which are not included here. The items which are assumed to be covered elsewhere are listed below:

- site supervision;

- site accommodation;

- lighting and power;

- water;

- safety, health and welfare;

- removal of rubbish;

- cleaning;

- drying;

- protection of work;

- security;

- insurances;

- scaffolding;

- temporary services;

- travelling time;

- temporary screens.

It should be noted that resources data on some specialist activities, such as lift installation and piling, are not included because of the difficulty in obtaining the information. Readers who need this data should contact The Lift Equipment Engineers Association (01279 816504) and The Federation of Piling Specialists (0181-663 0947) respectively.

The inclusion of BS numbers and Eurocodes has been kept to a minimum, whilst proprietory brand names have only been included where absolutely necessary. The reason for this decision lies in the aim of the book to provide estimators with information as broadly-based as possible. For example, the time taken to lay facing bricks is usually constant regardless of type, colour or cost. It is intended that this approach of providing data on a generic basis will enable the reader to extrapolate the data to similar materials or activities.

Detailed information has been included at the beginning of each chapter and is intended to assist the reader by presenting data relevant to the contents of the chapter. These include weights of materials and assessments of sundry items which must form part of the unit rate but are not mentioned in the item description, e.g. volumes of mortar per square metre of brickwork.

Labour

The labour constants are based upon work being carried out at floor or first floor level. The following adjustments should be made for working at other levels.

Basement	2.5%
2nd and 3rd floor	5%
4th and 5th floor	7.5%
6th and 7th floor	10%
8th and 9th floor	12.5%
10th and 11th floor	15%
12th and 13th floor	17.5%
14th and 15th floor	20%
16th and 17th floor	22.5%
18th and 19th floor	25%

There are 17 grades of labour in this book, as follows, and each chapter lists the relevant grades it contains:

LA	1 Craftsman
LB	1 Semi-skilled operative
LC	1 Unskilled operative

LD	2 Bricklayers and 1 unskilled operative
LE	1 Ganger and 1 unskilled operative
LF	2 Asphalters and 1 unskilled operative
LG	1 Roofer and 1 unskilled operative
LH	1 Ganger, 1 semi-skilled operative and 1 unskilled operative
LI	2 Craftsmen and 1 unskilled operative
LJ	1 Craftsman and 2 unskilled operatives
LK	1 Craftsman and 1 unskilled operative
LL	1 Ganger, 2 semi-skilled operatives and 1 unskilled operative
LM	1 Ganger, 1 craftsman, 2 semi-skilled operatives and 1 unskilled operative
LN	1 Ganger, 1 craftsman and 1 unskilled operative
LO	1 Ganger, 2 craftsmen and 1 unskilled operative
LP	1 Foreman, 1 advanced fitter/welder (gas/arc), 2 advanced fitter/welders (gas or arc), 3 advanced fitters, 2 fitters and 1 mate
LQ	1 Technician, 1 approved electrician, 1 electrician, 1 apprentice (18 year old) and 1 unskilled operative

Plant grades

There are 27 plant grades as follows.

PA	1 Hydraulic excavator (1.73m3)
PB	1 Compressor (375cfm)
PC	1 Skip (8m3)
PD	1 Tipper (6 wheel)
PE	1 Vibratory roller
PF	1 Concrete mixer (10/7)
PG	1 Asphalt boiler
PH	1 Tractor and harrow
PI	1 Tractor and seeder
PJ	1 Hand roller
PK	1 Hydraulic excavator (3.5m3)
PL	1 Hydraulic excavator (3.5m3) with compressor, drills and breakers
PM	1 Crawler dozer
PN	1 Mobile crane
PO	1 Tractor and roller
PP	1 Hydraulic excavator (1.7m3), 1 crawler dozer and 1 tipper (6 wheel)
PQ	1 Crawler crane and concrete skip

PR   1 Saw bench and 20% mobile crane

PS   1 Crawler crane

PT   1 Compressor and boiler

PU   1 Wheeled hydraulic excavator (1.7m3),
                      1 pump (170m3/h), 50 trench sheets,
                      50 props, 1 dumper (1.5t) and
                      1 vibratory compactor

PV   1 Crawler hydraulic excavator (1.7m3),
                      1 pump (275m3/h), 125 trench sheets,
                      100 props, 1 dumper (1.5t) and
                      1 vibratory compactor

PW   1 Wheeled hydraulic excavator (1.7m3),
                      1 dumper (1.5t) and 1 pump (170m3/h)

PX   1 Tractor and 1 motorised roller

PY   1 Concrete paver and 1 motorised roller

PZ   1 Tractor trailer and 1 crawler crane

PZA   1 Land drain trencher

Materials

The information provided on materials is intended to assist in the
ordering process. For example, the volume of concrete beds and
coverings to drain pipes is stated as well as the linear measurement.
Additional data on the weights of materials are also included as well as
other information intended to help the estimator.

Elemental costs

A chapter has been included on elemental costs for 20 different types of
buildings expressed in percentage terms. These percentages reflect the
apportionment of individual elementals and should help in the early
financial planning stages of a project.

PART ONE

BUILDING

1

Excavation and filling

Weights of materials	kg/m3
Ashes	800
Ballast	600
Chalk	2240
Clay	1800
Flint	2550
Gravel	1750
Hardcore	1900
Hoggin	1750
Sand	1600
Water	950

Shrinkage of deposited materials

Clay	- 10.0%
Gravel	- 7.5%
Sandy soil	- 12.5%

Bulking of excavated material

Clay	+ 40%
Gravel	+ 25%
Sand	+ 20%

Angle of repose	Type	Angle ^{0}
Earth	loose, dry	36-40
	loose, moist	45
	loose, wet	30
	consolidated, dry	42
	consolidated, moist	38
Loam	loose, dry	40-45
	loose, wet	20-25
Gravel	dry	35-45
	wet	25-30
Sand	loose, dry	35-40
	compact	30-35
	wet	25
Clay	loose, wet	20-25
	consolidated, moist	70

Labour grades

Craftsman	LA
Semi-skilled operative	LB
Unskilled operative	LC

Plant grades

Hydraulic excavator (1.7m3)	PA
Compressor (375cfm)	PB
Skip (8m3)	PC
Tipper wagon (6 wheel)	PD
Vibrating roller	PE
Tractor loader and vibrating roller	PO

Typical fuel consumption for plant

These figures relate to working in normal conditions. Reduce by 25% for light duties and increase by 50% for heavy duties.

Plant	Engine size kW	Litres/ hour
Compressors up to	20	4.0
	30	6.5
	40	8.2
	50	9.0
	75	16.0
	100	20.0
	125	25.0
	150	30.0
Concrete mixers up to	5	1.0
	10	2.4
	15	3.8
	20	5.0
Dumpers	5	1.3
	7	2.0
	10	3.0
	15	4.0
	20	4.9
	30	7.0
	50	12.0
Excavators	10	2.5
	20	4.5
	40	9.0
	60	13.0
	80	17.0
Pumps	5	1.1
	7.5	1.6
	10	2.1
	15	3.2
	20	4.2
	25	5.5

Plant	Engine size kW	Litres/ hour
Trenchers	25	5.0
	35	6.5
	50	10.0
	75	14.5

Average plant outputs (m3/hour)

Bucket size (litres)	Soil	Sand	Heavy clay	Soft rock
Face shovel				
200	11	12	7	5
300	18	20	12	9
400	24	26	17	13
600	42	45	28	23
Backactor				
200	8	8	6	4
300	12	13	9	7
400	17	18	11	10
600	28	30	19	15
Dragline				
200	11	12	8	5
300	18	20	12	9
400	25	27	16	12
600	42	45	28	21

	Unit	Labour grade	Labour hours	Plant grade	Plant hours	Materials

Site clearance

Cut down trees, grub
up roots

600 to 1500mm girth	nr	LC	8.00	PA	2.00	-
1500 to 3000mm girth	nr	LC	10.00	PA	4.00	-

Cut down hawthorn
hedge, grub up roots

1500mm high	m	LC	0.40	PA	0.20	-
3000mm high	m	LC	0.60	PA	0.30	-

Work by hand

Excavate topsoil, lay
aside for reuse, depth

150mm	m2	LC	0.40	-	-	-
200mm	m2	LC	0.50	-	-	-

Excavate from ground
level to reduce levels,
maximum depth not
exceeding

0.25m	m3	LC	1.80	-	-	-
1.00m	m3	LC	2.00	-	-	-
2.00m	m3	LC	4.00	-	-	-

Excavate pits to receive
bases, maximum depth
not exceeding

0.25m	m3	LC	2.75	-	-	-
1.00m	m3	LC	3.00	-	-	-
2.00m	m3	LC	3.25	-	-	-

	Unit	Labour grade	Labour hours	Plant grade	Plant hours	Materials

Excavate basements to receive bases, maximum depth not exceeding

1.00m	m3	LC	2.00	-	-	-
2.00m	m3	LC	2.60	-	-	-
4.00m	m3	LC	2.80	-	-	-
6.00m	m3	LC	3.50	-	-	-

Excavate trenches not exceeding 0.30m wide to receive foundations, maximum depth not exceeding

0.25m	m3	LC	2.50	-	-	-
1.00m	m3	LC	2.75	-	-	-
2.00m	m3	LC	3.00	-	-	-

Excavate trenches exceeding 0.30m wide to receive foundations, maximum depth not exceeding

0.25m	m3	LC	2.30	-	-	-
1.00m	m3	LC	2.50	-	-	-
2.00m	m3	LC	2.75	-	-	-

Excavate and fill working space to basement, depth not exceeding

0.25m	m2	LC	2.40	-	-	-
1.00m	m2	LC	2.50	-	-	-
2.00m	m2	LC	2.60	-	-	-

	Unit	Labour grade	Labour hours	Plant grade	Plant hours	Materials
Excavate and fill working space to pit, depth not exceeding						
0.25m	m2	LC	2.70	-	-	-
1.00m	m2	LC	2.80	-	-	-
2.00m	m2	LC	2.90	-	-	-
Excavate and fill working space to trenches, depth not exceeding						
0.25m	m2	LC	2.90	-	-	-
1.00m	m2	LC	3.00	-	-	-
2.00m	m2	LC	3.10	-	-	-
Extra for breaking up						
concrete 100mm thick	m2	LC	0.90	-	-	-
tarmacadam 75mm thick	m2	LC	0.50	-	-	-
hardcore 100mm thick	m2	LC	0.60	-	-	-
plain concrete	m3	LC	7.00	-	-	-
reinforced concrete	m3	LC	8.00	-	-	-
soft rock	m3	LC	10.00	-	-	-

Work by machine

	Unit	Labour grade	Labour hours	Plant grade	Plant hours	Materials
Excavate topsoil, lay aside for reuse, average depth						
150mm	m2	LC	0.02	PA	0.02	-
200mm	m2	LC	0.03	PA	0.03	-

	Unit	Labour grade	Labour hours	Plant grade	Plant hours	Materials
Excavate from ground level to reduce levels, maximum depth not exceeding						
0.25m	m3	LC	0.10	PA	0.04	-
1.00m	m3	LC	0.10	PA	0.04	-
2.00m	m3	LC	0.10	PA	0.04	-
Excavate pits to receive bases, maximum depth not exceeding						
0.25m	m3	LC	0.30	PA	0.10	-
1.00m	m3	LC	0.28	PA	0.08	-
2.00m	m3	LC	0.28	PA	0.08	-
Excavate basements to receive bases, maximum depth not exceeding						
1.00m	m3	LC	0.10	PA	0.04	-
2.00m	m3	LC	0.12	PA	0.05	-
4.00m	m3	LC	0.14	PA	0.06	-
6.00m	m3	LC	0.16	PA	0.07	-
Excavate trenches not exceeding 0.30m wide to receive foundations, maximum depth not exceeding						
0.25m	m3	LC	0.35	PA	0.08	-
1.00m	m3	LC	0.30	PA	0.06	-
2.00m	m3	LC	0.34	PA	0.07	-

	Unit	Labour grade	Labour hours	Plant grade	Plant hours	Materials

Excavate trenches exceeding 0.30m wide to receive foundations, maximum depth not exceeding

0.25m	m3	LC	0.30	PA	0.06	-
1.00m	m3	LC	0.25	PA	0.05	-
2.00m	m3	LC	0.30	PA	0.06	-

Excavate and fill working space to basement, depth not exceeding

0.25m	m2	LC	0.28	PA	0.06	-
1.00m	m2	LC	0.25	PA	0.05	-
2.00m	m2	LC	0.30	PA	0.07	-

Excavate and fill working space to pit, depth not exceeding

0.25m	m2	LC	0.24	PA	0.05	-
1.00m	m2	LC	0.20	PA	0.04	-
2.00m	m2	LC	0.28	PA	0.06	-

Excavate and fill working space to trench, depth not exceeding

0.25m	m2	LC	0.18	PA	0.04	-
1.00m	m2	LC	0.15	PA	0.03	-
2.00m	m2	LC	0.25	PA	0.05	-

	Unit	Labour grade	Labour hours	Plant grade	Plant hours	Materials
Extra for breaking up						
concrete 100mm thick	m2	LC	0.40	PB	0.40	-
tarmacadam 75mm thick	m2	LC	0.24	PB	0.24	-
hardcore 100mm thick	m2	LC	0.28	PB	0.28	-
plain concrete	m3	LC	3.00	PB	3.00	-
reinforced concrete	m3	LC	3.50	PB	3.50	-
soft rock	m3	LC	4.00	PB	4.00	-
hard rock	m2	LC	4.40	PB	4.40	-

Earthwork support

Earthwork support not exceeding 2m between opposing faces, depth not exceeding 1m in

	Unit	Labour grade	Labour hours	Plant grade	Plant hours	Materials
firm ground	m2	LA	0.45	-	-	-
loose ground	m2	LA	0.70	-	-	-
sand	m2	LA	0.90	-	-	-

Earthwork support not exceeding 2m between opposing faces, depth 2m in

	Unit	Labour grade	Labour hours	Plant grade	Plant hours	Materials
firm ground	m2	LA	0.50	-	-	-
loose ground	m2	LA	0.90	-	-	-
sand	m2	LA	1.10	-	-	-

Earthwork support not exceeding 2m between opposing faces, depth 3m in

	Unit	Labour grade	Labour hours	Plant grade	Plant hours	Materials
firm ground	m2	LA	0.52	-	-	-
loose ground	m2	LA	1.00	-	-	-
sand	m2	LA	1.25	-	-	-

	Unit	Labour grade	Labour hours	Plant grade	Plant hours	Materials
Earthwork support not exceeding 2m between opposing faces, depth 4m in						
firm ground	m2	LA	0.55	-	-	-
loose ground	m2	LA	1.20	-	-	-
sand	m2	LA	1.35	-	-	-
Earthwork support not exceeding 2m between opposing faces, depth 5m in						
firm ground	m2	LA	0.70	-	-	-
loose ground	m2	LA	1.35	-	-	-
sand	m2	LA	1.50	-	-	-
Earthwork support not exceeding 2m between opposing faces, depth 6m in						
firm ground	m2	LA	0.80	-	-	-
loose ground	m2	LA	0.50	-	-	-
sand	m2	LA	1.75	-	-	-
Earthwork support not exceeding 3m between opposing faces, depth not exceeding 1m in						
firm ground	m2	LA	0.50	-	-	-
loose ground	m2	LA	0.80	-	-	-
sand	m2	LA	1.00	-	-	-

	Unit	Labour grade	Labour hours	Plant grade	Plant hours	Materials
Earthwork support not exceeding 3m between opposing faces, depth 2m in						
firm ground	m2	LA	0.55	-	-	-
loose ground	m2	LA	1.00	-	-	-
sand	m2	LA	1.25	-	-	-
Earthwork support not exceeding 3m between opposing faces, depth 3m in						
firm ground	m2	LA	0.58	-	-	-
loose ground	m2	LA	1.20	-	-	-
sand	m2	LA	1.35	-	-	-
Earthwork support not exceeding 3m between opposing faces, depth 4m in						
firm ground	m2	LA	0.60	-	-	-
loose ground	m2	LA	1.35	-	-	-
sand	m2	LA	1.50	-	-	-
Earthwork support not exceeding 3m between opposing faces, depth 5m in						
firm ground	m2	LA	0.75	-	-	-
loose ground	m2	LA	1.50	-	-	-
sand	m2	LA	1.70	-	-	-

	Unit	Labour grade	Labour hours	Plant grade	Plant hours	Materials

Earthwork support not exceeding 3m between opposing faces, depth 6m in

	Unit	Labour grade	Labour hours	Plant grade	Plant hours	Materials
firm ground	m2	LA	0.90	-	-	-
loose ground	m2	LA	1.65	-	-	-
sand	m2	LA	1.90	-	-	-

Earthwork support not exceeding 4m between opposing faces, depth 1m in

	Unit	Labour grade	Labour hours	Plant grade	Plant hours	Materials
firm ground	m2	LA	0.55	-	-	-
loose ground	m2	LA	0.90	-	-	-
sand	m2	LA	1.10	-	-	-

Earthwork support not exceeding 4m between opposing faces, depth 2m in

	Unit	Labour grade	Labour hours	Plant grade	Plant hours	Materials
firm ground	m2	LA	0.60	-	-	-
loose ground	m2	LA	1.10	-	-	-
sand	m2	LA	1.35	-	-	-

Earthwork support not exceeding 4m between opposing faces, depth 3m in

	Unit	Labour grade	Labour hours	Plant grade	Plant hours	Materials
firm ground	m2	LA	0.62	-	-	-
loose ground	m2	LA	1.35	-	-	-
sand	m2	LA	1.50	-	-	-

	Unit	Labour grade	Labour hours	Plant grade	Plant hours	Materials

Earthwork support not exceeding 4m between opposing faces, depth 4m in

	Unit	Labour grade	Labour hours	Plant grade	Plant hours	Materials
firm ground	m2	LA	0.65	-	-	-
loose ground	m2	LA	1.50	-	-	-
sand	m2	LA	1.65	-	-	-

Earthwork support not exceeding 4m between opposing faces, depth 5m in

	Unit	Labour grade	Labour hours	Plant grade	Plant hours	Materials
firm ground	m2	LA	0.80	-	-	-
loose ground	m2	LA	1.75	-	-	-
sand	m2	LA	2.10	-	-	-

Earthwork support not exceeding 4m between opposing faces, depth 6m in

	Unit	Labour grade	Labour hours	Plant grade	Plant hours	Materials
firm ground	m2	LA	1.00	-	-	-
loose ground	m2	LA	2.20	-	-	-
sand	m2	LA	2.75	-	-	-

Disposal

Load surplus excavated material into barrows, wheel and deposit in temporary spoil heaps, average distance

	Unit	Labour grade	Labour hours	Plant grade	Plant hours	Materials
25m	m3	LC	1.20	-	-	-
50m	m3	LC	1.50	-	-	-

	Unit	Labour grade	Labour hours	Plant grade	Plant hours	Materials tonnes
Load into barrows, wheel and deposit in skip or lorry, average distance 25m	m3	LC	1.40	-	-	-

Remove from site by
lorry including tipping
charges, average distance

5km	m3	-	-	PD	0.15	-
8km	m3	-	-	PD	0.20	-
10km	m3	-	-	PD	0.25	-
12km	m3	-	-	PD	0.30	-
20km	m3	-	-	PD	0.45	-

Filling

Filling material deposited
on site in layers not
exceeding 250mm thick,
average distance 25m

surplus excavated material	m3	LC	0.33	PO	0.10	1.80
sand	m3	LC	0.42	PO	0.10	1.60
hardcore	m3	LC	0.42	PO	0.10	1.90
granular fill	m3	LC	0.42	PO	0.10	1.90
imported soil	m3	LC	0.42	PO	0.10	1.60

Filling material deposited
on site average distance
25m, in layer 100mm thick

surplus excavated material	m2	LC	0.04	PO	0.01	0.18
sand	m2	LC	0.05	PO	0.01	0.16
hardcore	m2	LC	0.05	PO	0.01	0.19
granular fill	m2	LC	0.05	PO	0.01	0.19
imported soil	m2	LC	0.05	PO	0.01	0.16

	Unit	Labour grade	Labour hours	Plant grade	Plant hours	Materials tonnes

Filling material deposited
on site average distance 25m,
in layer 150mm thick

surplus excavated

	Unit	Labour grade	Labour hours	Plant grade	Plant hours	Materials tonnes
material	m2	LC	0.06	PO	0.01	0.27
sand	m2	LC	0.07	PO	0.01	0.24
hardcore	m2	LC	0.07	PO	0.01	0.28
granular fill	m2	LC	0.07	PO	0.01	0.28
imported soil	m2	LC	0.07	PO	0.01	0.24

Filling material deposited
on site average distance 25m,
in layer 200mm thick

surplus excavated

	Unit	Labour grade	Labour hours	Plant grade	Plant hours	Materials tonnes
material	m2	LC	0.08	PO	0.02	0.36
sand	m2	LC	0.10	PO	0.02	0.32
hardcore	m2	LC	0.10	PO	0.02	0.38
granular fill	m2	LC	0.10	PO	0.02	0.38
imported soil	m2	LC	0.10	PO	0.02	0.32

Filling material deposited
on site average distance 25m,
in layer 225mm thick

surplus excavated

	Unit	Labour grade	Labour hours	Plant grade	Plant hours	Materials tonnes
material	m2	LC	0.12	PO	0.02	0.41
sand	m2	LC	0.12	PO	0.02	0.36
hardcore	m2	LC	0.12	PO	0.02	0.42
granular fill	m2	LC	0.12	PO	0.02	0.42
imported soil	m2	LC	0.12	PO	0.02	0.36

	Unit	Labour grade	Labour hours	Plant grade	Plant hours	Materials tonnes

Surface treatments

Level and compact
excavation with
vibrating roller

	Unit	Labour grade	Labour hours	Plant grade	Plant hours	Materials tonnes
surplus excavated material	m2	LC	0.12	PE	0.03	-
sand	m2	LC	0.11	PE	0.03	-
hardcore	m2	LC	0.12	PE	0.03	-

2

In situ and precast concrete

Weights of materials	kg/m3
Cement	1440
Sand	1600
Aggregate, coarse	1500
Stone, crushed	1350
Ballast, all-in	1800
Concrete	2450

Suitability of mixes

Precast work in small sectional areas	1:1:2
Watertight reinforced concrete structures	1:1.5:3
Normal reinforced concrete work	1:2:4
Mass unreinforced concrete work	1:2.5:5
Rough concrete work	1:3:6

Concrete mixes (per m3)

Mix	Cement t	Sand m3	Aggregate m3	Water litres
1:1:2	0.50	0.45	0.70	208
1:1.5:3	0.37	0.50	0.80	185
1:2:4	0.30	0.54	0.85	175

Mix	Cement t	Sand m3	Aggregate m3	Water litres
1:2.5:5	0.25	0.55	0.85	166
1:3:6	0.22	0.55	0.85	160
Grade				
20/20	0.32	0.62	1.20	170
25/20	0.35	0.60	1.17	180
30/20	0.80	0.59	1.11	200
7/40 all-in	0.18	-	1.95	150
20/20 all-in	0.32	-	1.85	170
25/20 all-in	0.36	-	1.75	180

Steel bar reinforcement

Diameter mm	Nominal weight kg/m	Length m/tonne	Sectional area mm2
6	0.222	4505	28.30
8	0.395	2532	50.30
10	0.616	1623	78.50
12	0.888	1126	113.10
16	1.579	633	201.10
20	2.466	406	314.20
25	3.854	259	490.90
32	6.313	158	804.20
40	9.864	101	1256.60
50	15.413	65	1963.50

Steel fabric reinforcement

BS4483 ref.	Nominal weight kg/m^2	Mesh dimensions Main mm	Cross mm	Wire diameters Main mm	Cross mm
A393	6.16	200	200	10	10
A252	3.95	200	200	8	8
A193	3.02	200	200	7	7
A142	2.22	200	200	6	6
A98	1.54	200	200	5	5
B1131	10.90	100	200	12	8
B785	8.14	100	200	10	8
B503	5.93	100	200	8	8
B385	4.53	100	200	7	7
B283	3.73	100	200	6	7
B196	3.05	100	200	5	7
C785	6.72	100	400	10	6
C636	5.55	100	400	9	6
C503	4.34	100	400	8	5
C385	3.41	100	400	7	5
C283	2.61	100	400	6	5
D98	1.54	200	200	5	5

Formwork stripping times

	Ordinary concrete about 60°	about 35°	Rapid hardening concrete about 60°	about 35°
Beams, columns, walls	1	1	6	5
Soffits of slabs	3	10	2	7
Soffits of beams	7	12	4	10

Labour grades

Craftsman	LA
Semi-skilled operative	LB

Plant grades

Concrete mixer (10/7)	PF

	Unit	Labour grade	Labour hours	Plant grade	Plant hours

Ready mixed concrete

Mix 1:3:6 (11.50N/mm2 40mm aggregate)

	Unit	Labour grade	Labour hours	Plant grade	Plant hours
Foundations in trenches	m3	LB	1.35	-	-
Isolated bases	m3	LB	1.90	-	-
Beds					
over 450mm thick	m3	LB	1.35	-	-
150 to 450mm thick	m3	LB	1.70	-	-
not exceeding 150mm thick	m3	LB	2.80	-	-
Extra for placing concrete around reinforcement	m3	LB	0.55	-	-

Mix 1:2:4 (21.00N/mm2 aggregate)

	Unit	Labour grade	Labour hours	Plant grade	Plant hours
Isolated bases	m3	LB	1.35	-	-
Beds					
over 450mm thick	m3	LB	1.35	-	-
150 to 450mm thick	m3	LB	1.70	-	-
not exceeding 150mm thick	m3	LB	2.80	-	-
Suspended slabs					
over 450mm thick	m3	LB	1.35	-	-
150 to 450mm thick	m3	LB	1.70	-	-
not exceeding 150mm thick	m3	LB	2.80	-	-

Mix 1:2:4 (21.00N/mm2 aggregate)

	Unit	Labour grade	Labour hours	Plant grade	Plant hours
Walls					
over 450mm thick	m3	LB	2.60	-	-
150 to 450mm thick	m3	LB	3.60	-	-
not exceeding 150mm thick	m3	LB	4.20	-	-

	Unit	Labour grade	Labour hours	Plant grade	Plant hours
Casings to isolated beam	m3	LB	5.00	-	-
Casings to isolated deep beam	m3	LB	4.50	-	-
Casings to attached deep beam	m3	LB	4.50	-	-
Isolated beam	m3	LB	5.00	-	-
Isolated deep beam	m3	LB	4.50	-	-
Attached deep beam	m3	LB	4.50	-	-
Columns	m3	LB	5.00	-	-
Column casings	m3	LB	5.00	-	-
Staircases	m3	LB	6.00	-	-

Site mixed concrete

Mix 1:3:6 (11.50N/mm2, 40mm aggregate)

	Unit	Labour grade	Labour hours	Plant grade	Plant hours
Foundations in trenches	m3	LB	1.95	PF	0.50
Isolated bases	m3	LB	2.85	PF	0.50

Mix 1:2:4 (21.00N/mm2, 20mm aggregate)

	Unit	Labour grade	Labour hours	Plant grade	Plant hours
Isolated bases	m3	LB	2.85	PF	0.50

Mix 1:2:4 (21.00N/mm2, 20mm aggregate)

Beds

	Unit	Labour grade	Labour hours	Plant grade	Plant hours
over 450mm thick	m3	LB	1.95	PF	0.50
150 to 450mm thick	m3	LB	2.55	PF	0.50
not exceeding 150mm thick	m3	LB	3.70	PF	0.50

	Unit	Labour grade	Labour hours	Plant grade	Plant hours
Suspended slabs					
over 450mm thick	m3	LB	1.95	PF	0.50
150 to 450mm thick	m3	LB	2.55	PF	0.50
not exceeding 150mm thick	m3	LB	3.70	PF	0.50
Extra for placing around reinforcement	m3	LB	0.55	-	-
Extra for laying to slopes not exceeding 15 degrees	m2	LB	0.55	-	-
Extra for laying to slopes over 15 degrees	m2	LB	0.55	-	-
Formwork					
Sides of foundations, bases, beams or beds, height					
over 1m	m2	LA	1.80	-	-
not exceeding 250mm	m	LA	0.60	-	-
250 to 500mm	m	LA	0.90	-	-
500mm to 1m	m	LA	1.40	-	-
Sides of ground beams and edges of beds					
over 1m	m2	LA	1.80	-	-
not exceeding 250mm	m	LA	0.60	-	-
250 to 500mm	m	LA	0.90	-	-
500mm to 1m	m	LA	1.40	-	-
Edges of suspended slabs					
not exceeding 250mm	m	LA	0.90	-	-
250 to 500mm	m	LA	1.40	-	-
500mm to 1m	m	LA	1.80	-	-

	Unit	Labour grade	Labour hours	Plant grade	Plant hours
Sides of upstands					
over 1m	m2	LA	2.00	-	-
not exceeding 250mm	m	LA	0.80	-	-
250 to 500mm	m	LA	1.00	-	-
500mm to 1m	m	LA	1.80	-	-
Steps in top surfaces					
not exceeding 250mm	m	LA	0.60	-	-
250 to 500mm	m	LA	1.00	-	-
500mm to 1m	m	LA	1.20	-	-
Steps in soffits					
not exceeding 250mm	m	LA	0.70	-	-
250 to 500mm	m	LA	1.10	-	-
500mm to 1m	m	LA	1.30	-	-
Machine bases and plinths					
over 1m	m2	LA	1.80	-	-
not exceeding 250mm	m	LA	0.60	-	-
250 to 500mm	m	LA	1.90	-	-
500mm to 1m	m	LA	1.40	-	-
Soffits of slabs					
not exceeding 250mm	m2	LA	1.80	-	-
250 to 500mm	m2	LA	1.85	-	-
500mm to 1m	m2	LA	1.90	-	-
Walls					
vertical	m2	LA	1.80	-	-
vertical, interrupted	m2	LA	2.10	-	-
vertical, exceeding 3m	m2	LA	2.00	-	-

	Unit	Labour grade	Labour hours	Plant grade	Plant hours

Beams, rectangular attached to in situ slabs, girth of beam

	Unit	Labour grade	Labour hours	Plant grade	Plant hours
500mm	m2	LA	1.85	-	-
600mm	m2	LA	1.90	-	-
700mm	m2	LA	2.05	-	-
800mm	m2	LA	2.15	-	-
900mm	m2	LA	2.25	-	-
1000mm	m2	LA	2.40	-	-
1100mm	m2	LA	2.60	-	-
1200mm	m2	LA	2.75	-	-

Isolated columns, girth

	Unit	Labour grade	Labour hours	Plant grade	Plant hours
500mm	m2	LA	1.75	-	-
600mm	m2	LA	1.80	-	-
700mm	m2	LA	1.95	-	-
800mm	m2	LA	2.05	-	-
900mm	m2	LA	2.10	-	-
1000mm	m2	LA	2.25	-	-
1100mm	m2	LA	2.40	-	-
1200mm	m2	LA	2.50	-	-

Columns, rectangular attached to in situ walls, girth of column

	Unit	Labour grade	Labour hours	Plant grade	Plant hours
500mm	m2	LA	1.65	-	-
600mm	m2	LA	1.75	-	-
700mm	m2	LA	1.85	-	-
800mm	m2	LA	1.95	-	-
900mm	m2	LA	2.05	-	-
1000mm	m2	LA	2.15	-	-
1100mm	m2	LA	2.30	-	-
1200mm	m2	LA	2.40	-	-

	Unit	Labour grade	Labour hours	Plant grade	Plant hours

Sundries

Mortice in concrete for rag bolt, grout in cement mortar (1:1), depth

50mm	nr	LA	0.25	-	-
100mm	nr	LA	0.30	-	-
150mm	nr	LA	0.35	-	-

Mortice in concrete for holding down bolt and plates, grout in cement mortar (1:3), depth

200mm	nr	LA	0.60	-	-
400mm	nr	LA	0.80	-	-

Cut chase not exceeding 50mm deep in concrete, width

25mm	m	LA	0.30	-	-
50mm	m	LA	0.60	-	-
75mm	m	LA	0.80	-	-

Cut chase not exceeding 100mm deep in concrete, width

50mm	m	LA	1.00	-	-
75mm	m	LA	1.10	-	-
100mm	m	LA	1.20	-	-

Cut chase not exceeding 150mm deep in concrete, width

100mm	m	LA	1.50	-	-
150mm	m	LA	1.75	-	-

	Unit	Labour grade	Labour hours	Plant grade	Plant hours
Cut chase not exceeding 50mm deep in reinforced concrete, width					
25mm	m	LA	0.50	-	-
50mm	m	LA	0.90	-	-
75mm	m	LA	1.20	-	-
Cut chase not exceeding 100mm deep in reinforced concrete, width					
50mm	m	LA	1.40	-	-
75mm	m	LA	1.50	-	-
100mm	m	LA	1.60	-	-
Cut chase not exceeding 150mm deep in concrete, width					
100mm	m	LA	2.00	-	-
150mm	m	LA	2.20	-	-
Cut hole in reinforced concrete not exceeding 100mm thick, size					
150 x 150mm	nr	LA	0.55	-	-
200 x 200mm	nr	LA	0.65	-	-
300 x 200mm	nr	LA	0.75	-	-
25mm diameter	nr	LA	0.30	-	-
50mm diameter	nr	LA	0.40	-	-
100mm diameter	nr	LA	0.45	-	-

	Unit	Labour grade	Labour hours	Plant grade	Plant hours
Cut hole in reinforced concrete not exceeding 150mm thick, size					
150 x 150mm	nr	LA	0.75	-	-
200 x 200mm	nr	LA	0.85	-	-
300 x 200mm	nr	LA	0.95	-	-
25mm diameter	nr	LA	0.40	-	-
50mm diameter	nr	LA	0.50	-	-
100mm diameter	nr	LA	0.55	-	-
Cut hole in reinforced concrete not exceeding 200mm thick, size					
150 x 150mm	nr	LA	0.90	-	-
200 x 200mm	nr	LA	1.00	-	-
300 x 200mm	nr	LA	1.10	-	-
25mm diameter	nr	LA	0.50	-	-
50mm diameter	nr	LA	0.55	-	-
100mm diameter	nr	LA	0.65	-	-
Cut hole in reinforced concrete not exceeding 300mm thick, size					
150 x 150mm	nr	LA	1.25	-	-
200 x 200mm	nr	LA	1.50	-	-
Cut hole in reinforced concrete not exceeding 300mm thick, size					
300 x 200mm	nr	LA	1.70	-	-
25mm diameter	nr	LA	0.80	-	-
50mm diameter	nr	LA	1.00	-	-
100mm diameter	nr	LA	1.25	-	-

	Unit	Labour grade	Labour hours	Plant grade	Plant hours

Reinforcement

Reinforcement bars, plain round steel, straight or bent

	Unit	Labour grade	Labour hours	Plant grade	Plant hours
6mm	t	LA	80.00	-	-
8mm	t	LA	70.00	-	-
10mm	t	LA	60.00	-	-
12mm	t	LA	50.00	-	-
16mm	t	LA	40.00	-	-
20mm	t	LA	30.00	-	-
25mm	t	LA	25.00	-	-
32mm	t	LA	22.00	-	-
40mm	t	LA	20.00	-	-

Steel fabric reinforcement, 200mm laps, laid in concrete beds

Ref.	kg/m2	Unit	Labour grade	Labour hours	Plant grade	Plant hours
A98	1.54	m2	LA	0.11	-	-
A142	2.22	m2	LA	0.12	-	-
A193	3.02	m2	LA	0.15	-	-
A252	3.95	m2	LA	0.17	-	-
A393	6.16	m2	LA	0.20	-	-
B196	3.05	m2	LA	0.15	-	-

		Unit	Labour grade	Labour hours	Plant grade	Plant hours
B283	3.73	m2	LA	0.16	-	-
B385	4.53	m2	LA	0.17	-	-
B503	5.93	m2	LA	0.19	-	-
B785	8.14	m2	LA	0.22	-	-
B1131	10.90	m2	LA	0.24	-	-
C283	2.61	m2	LA	0.13	-	-
C385	3.41	m2	LA	0.16	-	-
C503	4.34	m2	LA	0.18	-	-
C636	5.55	m2	LA	0.19	-	-
C785	6.72	m2	LA	0.21	-	-

Cutting on mesh reinforcement

	Unit	Labour grade	Labour hours	Plant grade	Plant hours
raking	m2	LA	0.10	-	-
curved	m2	LA	0.20	-	-

Designed joints

Expansion joint, impregnated
fibre based joint filler, formed joint

12.5mm thick

	Unit	Labour grade	Labour hours	Plant grade	Plant hours
not exceeding 150mm wide	m	LB	0.12	-	-
150 to 300mm wide	m	LB	0.18	-	-
300 to 450mm wide	m	LB	0.20	-	-

20mm thick

	Unit	Labour grade	Labour hours	Plant grade	Plant hours
not exceeding 150mm wide	m	LB	0.15	-	-
150 to 300mm wide	m	LB	0.20	-	-
300 to 450mm wide	m	LB	0.25	-	-

	Unit	Labour grade	Labour hours	Materials

Waterstops

Flat PVC dumbbell waterstop with
welded joints in formed joint, width

100mm	m	LB	0.25	-
170mm	m	LB	0.28	-
210mm	m	LB	0.34	-
250mm	m	LB	0.40	-

Flat angle to PVC dumbbell
waterstop, width

100mm	nr	LB	0.25	-
170mm	nr	LB	0.28	-
210mm	nr	LB	0.34	-
250mm	nr	LB	0.40	-

Vertical angle to PVC
dumbbell waterstop, width

100mm	nr	LB	0.25	-
170mm	nr	LB	0.28	-
210mm	nr	LB	0.34	-
250mm	nr	LB	0.40	-

3-way flat intersection to
PVC dumbbell waterstop, width

100mm	nr	LB	0.30	-
170mm	nr	LB	0.33	-
210mm	nr	LB	0.35	-
250mm	nr	LB	0.45	-

3-way vertical intersection to
PVC dumbbell waterstop, width

100mm	nr	LB	0.30	-
170mm	nr	LB	0.33	-
210mm	nr	LB	0.35	-
250mm	nr	LB	0.45	-

	Unit	Labour grade	Labour hours	Materials
4-way flat intersection to PVC dumbbell waterstop, width				
100mm	nr	LB	0.38	-
170mm	nr	LB	0.41	-
210mm	nr	LB	0.46	-
250mm	nr	LB	0.53	-
Flat PVC centre bulb waterstop with welded joints in formed joint, width				
100mm	m	LB	0.26	-
170mm	m	LB	0.30	-
Flat PVC centre bulb waterstop with welded joints in formed joint, width				
210mm	m	LB	0.35	-
250mm	m	LB	0.38	-
Flat angle to PVC centre bulb waterstop, width				
100mm	nr	LB	0.26	-
170mm	nr	LB	0.30	-
210mm	nr	LB	0.35	
250mm	nr	LB	0.38	-
Vertical angle to PVC centre bulb waterstop, width				
100mm	nr	LB	0.26	-
170mm	nr	LB	0.30	-
210mm	nr	LB	0.35	-
250mm	nr	LB	0.38	-

	Unit	Labour grade	Labour hours	Materials
3-way flat intersection to PVC centre stop waterstop, width				
100mm	nr	LB	0.31	-
170mm	nr	LB	0.35	-
210mm	nr	LB	0.40	-
250mm	nr	LB	0.43	-
3-way vertical intersection to PVC centre stop waterstop, width				
100mm	nr	LB	0.31	-
170mm	nr	LB	0.35	-
210mm	nr	LB	0.40	-
250mm	nr	LB	0.43	-
4-way flat intersection to PVC dumbbell waterstop, width				
100mm	nr	LB	0.39	-
170mm	nr	LB	0.43	-
210mm	nr	LB	0.48	-
250mm	nr	LB	0.51	-

Worked finishes on in situ concrete

Prepare level surfaces of unset concrete

tamping by mechanical means	m2	LB	0.06	-
power floating	m2	LB	0.15	-
trowelling	m2	LB	0.15	-

	Unit	Labour grade	Labour hours	Mortar m3

Precast concrete

Copings, once weathered
and once throated

150 x 75mm	m	LC	0.35	0.002
300 x 75mm	m	LC	0.40	0.004

Copings, twice weathered
and twice throated

150 x 75mm	m	LC	0.40	0.002
300 x 75mm	m	LC	0.45	0.004

Sills, once weathered
and once throated

150 x 65 x 900mm	nr	LC	0.35	0.002
150 x 65 x 1200mm	nr	LC	0.40	0.003
150 x 65 x 1500mm	nr	LC	0.45	0.004
200 x 75 x 900mm	nr	LC	0.45	0.004
200 x 75 x 1200mm	nr	LC	0.50	0.005
200 x 75 x 1500mm	nr	LC	0.55	0.006

Lintels, rectangular

75 x 150 x 900mm	nr	LC	0.40	0.002
75 x 150 x 1200mm	nr	LC	0.45	0.003
75 x 150 x 1500mm	nr	LC	0.50	0.004
100 x 225 x 900mm	nr	LC	0.45	0.004
100 x 225 x 1200mm	nr	LC	0.50	0.005
100 x 225 x 1500mm	nr	LC	0.55	0.006

Pier cappings

300 x 300 x 75mm	nr	LC	0.25	0.001
400 x 400 x 75mm	nr	LC	0.30	0.001
450 x 450 x 75mm	nr	LC	0.35	0.001
500 x 500 x 75mm	nr	LC	0.40	0.002

3

Brickwork and blockwork

Weights of materials

Cement	1440kg/m3
Sand	1600kg/m3
Lime, ground	750kg/m3
Brickwork, 112.5mm	220kg/m2
215 mm	465kg/m2
327.5mm	710kg/m2
Stone, natural	2400kg/m3
reconstructed	2250kg/m3
Bricks, Fletton	1820kg/m2
engineering	2250kg/m2
concrete	1850kg/m2
Blocks, natural aggregate	
75mm thick	160kg/m2
100mm thick	215kg/m2
140mm thick	300kg/m2

Blocks, lightweight aggregate

75mm thick	60kg/m2
100mm thick	80kg/m2
140mm thick	112kg/m2

Bricks per m2 (brick size 215 x 103.5 x 65mm)

Half brick wall

stretcher bond	59
English bond	89
English garden wall bond	74
Flemish bond	79

One brick wall

English bond	118
Flemish bond	118

One and a half brick wall

English bond	178
Flemish bond	178

Two brick wall

English bond	238
Flemish bond	238

Metric modular bricks

200 x 100 x 75mm	67
90mm thick	133
190mm thick	200
200 x 100 x 100mm	
90mm thick	50
190mm thick	100
290mm thick	150
300 x 100 x 75mm	
90mm thick	44
300 x 100 x 100mm	
90mm thick	33

Blocks per m2 (block size 414 x 215mm)

60mm thick	9.9
75mm thick	9.9
100mm thick	9.9
140mm thick	9.9
190mm thick	9.9
215mm thick	9.9

Mortar per m2

Brick size 215 x 103.5 x 65mm	Wirecut m3	1 frog m3	2 frogs m3
Half brick wall	0.017	0.024	0.031
One brick wall	0.045	0.059	0.073
One and a half brick wall	0.072	0.093	0.114
Two brick wall	0.101	0.128	0.155

Brick size 200 x 100 x 75mm

	Solid	Perforated
90mm thick	0.016	0.019
190mm thick	0.042	0.048
290mm thick	0.068	0.078

Brick size 200 x 100 x 100mm

	Solid	Perforated
90mm thick	0.013	0.016
190mm thick	0.036	0.041
290mm thick	0.059	0.067

Brick size 300 x 100 x 75mm

	Solid	Perforated
90mm thick	0.015	0.018

Brick size 300 x 100 x 100mm

	Solid	Perforated
90mm thick		0.015

Block size 440 x 215mm

60mm	0.004
75mm	0.005
100mm	0.006
140mm	0.007
190mm	0.008
215mm	0.009

Length of pointing per m2 (one face only)

English bond	19.1m
English garden wall bond	18.1m
Flemish bond	18.4m
Flemish garden wall bond	17.7m

Damp-proof courses	**kg/m2**
Hessian base	3.8
Fibre base	3.3
Asbestos base	3.8
Hessian base and lead core	4.4
Asbestos base and lead core	4.9
Pitch polymer	4.8

Labour grades

Craftsman   LA

2 Bricklayers and 1 unskilled
operative   LD

	Unit	Labour grade	Labour hours	Bricks nr	Mortar m3

Brickwork

Common bricks in gauged mortar (1:3)

Walls, facework one side

half brick wall	m2	LD	0.70	59	0.017
one brick wall	m2	LD	1.10	118	0.045
one and a half brick wall	m2	LD	1.50	178	0.072
two brick wall	m2	LD	2.10	238	0.101

Walls, facework two sides

half brick wall	m2	LD	0.80	59	0.017
one brick wall	m2	LD	1.20	118	0.045
one and a half brick wall	m2	LD	1.60	178	0.072
two brick wall	m2	LD	2.20	238	0.101

Skins of hollow walls

half brick wall	m2	LD	0.80	59	0.017
one brick wall	m2	LD	1.20	118	0.045

Honeycombed walls

half brick wall	m2	LD	0.60	38	0.011

Dwarf solid walls

half brick wall	m2	LD	0.60	59	0.017

	Unit	Labour grade	Labour hours	Bricks nr	Mortar m3
Isolated casings					
one brick wall	m2	LD	1.40	118	0.045
one and a half brick wall	m2	LD	1.80	178	0.072
two brick wall	m2	LD	2.40	238	0.101
Chimney stacks					
one brick wall	m2	LD	1.80	118	0.045
one and a half brick wall	m2	LD	2.20	178	0.072
two brick wall	m2	LD	2.80	238	0.101
Backing to masonry, cutting and bonding					
one brick wall	m2	LD	1.40	118	0.045
one and a half brick wall	m2	LD	2.40	178	0.072
Projections of chimney breasts					
half brick wall	m2	LD	1.00	59	0.017
one brick wall	m2	LD	1.40	118	0.045
one and a half brick wall	m2	LD	2.10	178	0.072
two brick wall	m2	LD	2.40	238	0.101
Projections of attached piers, plinths, bands and the like					
225 x 112mm	m	LD	0.55	13	0.006
225 x 225mm	m	LD	1.47	26	0.011
337 x 225mm	m	LD	2.10	39	0.017

	Unit	Labour grade	Labour hours	Bricks nr	Mortar m3
Bonding ends to existing					
half brick wall	m	LD	0.21	-	0.017
one brick wall	m	LD	0.50	-	0.045
one and a half brick wall	m	LD	0.75	-	0.072
two brick wall	m	LD	1.10	-	0.101

Class B engineering bricks in cement mortar (1:3)

Walls, facework one side					
half brick wall	m2	LD	0.80	59	0.017
one brick wall	m2	LD	1.20	118	0.045
one and a half brick wall	m2	LD	1.60	178	0.072
two brick wall	m2	LD	2.20	238	0.101

Walls, facework two sides					
half brick wall	m2	LD	0.90	59	0.017
one brick wall	m2	LD	1.30	118	0.045
one and a half brick wall	m2	LD	1.70	178	0.072
two brick wall	m2	LD	2.30	238	0.101

Skins of hollow walls					
half brick wall	m2	LD	0.90	59	0.017
one brick wall	m2	LD	1.30	118	0.045

Isolated casings					
one brick wall	m2	LD	1.50	118	0.045
one and a half brick wall	m2	LD	1.90	178	0.072
two brick wall	m2	LD	2.50	238	0.101

	Unit	Labour grade	Labour hours	Bricks nr	Mortar m3
Chimney stacks					
one brick wall	m2	LD	1.90	118	0.045
one and a half brick wall	m2	LD	2.30	178	0.072
two brick wall	m2	LD	2.90	238	0.101
Backing to masonry, cutting and bonding					
one brick wall	m2	LD	1.50	118	0.045
one and a half brick wall	m2	LD	1.90	178	0.072
Projections of chimney breasts					
half brick wall	m2	LD	1.10	59	0.017
one brick wall	m2	LD	1.40	118	0.045
one and a half brick wall	m2	LD	2.10	178	0.072
two brick wall	m2	LD	2.40	238	0.101
Projections of attached piers, plinths, bands and the like					
225 x 112mm	m	LD	0.71	13	0.006
225 x 225mm	m	LD	1.70	26	0.011
337 x 225mm	m	LD	2.32	39	0.017
Bonding ends to existing					
half brick wall	m	LD	0.35	-	0.017
one brick wall	m	LD	0.50	-	0.045
one and a half brick wall	m	LD	0.75	-	0.072
two brick wall	m	LD	1.10	-	0.101

	Unit	Labour grade	Labour hours	Bricks nr	Mortar m3

Facing bricks in gauged mortar (1:3)

Walls, facework one side

half brick wall	m2	LD	0.90	59	0.017
half brick in hollow wall	m2	LD	1.00	59	0.017
one brick wall	m2	LD	1.30	118	0.045

Walls, facework two sides

half brick wall	m2	LD	1.00	59	0.017
one brick wall	m2	LD	1.40	118	0.045

Bonding ends to existing

half brick wall	m	LD	0.35	-	0.017
one brick wall	m	LD	0.50	-	0.045
one and a half brick wall	m	LD	0.75	-	0.072
two brick wall	m	LD	1.10	-	0.101

Blockwork

Precast concrete natural aggregate block in gauged mortar (1:1:6)

In walls and partitions, thickness

75mm	m2	LD	0.45	9.90	0.005
100mm	m2	LD	0.50	9.90	0.006
140mm	m2	LD	0.60	9.90	0.007
215mm	m2	LD	0.70	9.90	0.007

	Unit	Labour grade	Labour hours	Blocks nr	Mortar m3
In skins of hollow walls, thickness					
75mm	m2	LD	0.55	9.90	0.005
100mm	m2	LD	0.60	9.90	0.006
140mm	m2	LD	0.70	9.90	0.007
215mm	m2	LD	0.80	9.90	0.007
In piers and chimney breasts, thickness					
75mm	m2	LD	0.75	9.90	0.005
100mm	m2	LD	0.80	9.90	0.006
140mm	m2	LD	0.90	9.90	0.007
215mm	m2	LD	1.00	9.90	0.007
In isolated casings, thickness					
75mm	m2	LD	0.95	9.90	0.005
100mm	m2	LD	1.00	9.90	0.006
140mm	m2	LD	1.10	9.90	0.007
215mm	m2	LD	1.20	9.90	0.007
Extra for fair face and flush pointing					
one side	m2	LD	0.16	-	-
two sides	m2	LD	0.33	-	-
Bonding ends of blockwork to brickwork in alternate courses					
75mm	m2	LD	0.30	-	-
100mm	m2	LD	0.38	-	-
175mm	m2	LD	0.44	-	-
215mm	m2	LD	0.62	-	-

	Unit	Labour grade	Labour hours	Blocks nr	Mortar m3
Precast concrete clinker aggregate block in gauged mortar (1:1:6)					
In walls and partitions, thickness					
60mm	m2	LD	0.40	9.90	0.004
75mm	m2	LD	0.45	9.90	0.005
100mm	m2	LD	0.50	9.90	0.006
140mm	m2	LD	0.60	9.90	0.007
215mm	m2	LD	0.70	9.90	0.007
In skins of hollow walls, thickness					
60mm	m2	LD	0.50	9.90	0.004
75mm	m2	LD	0.55	9.90	0.005
100mm	m2	LD	0.60	9.90	0.006
140mm	m2	LD	0.70	9.90	0.007
215mm	m2	LD	0.80	9.90	0.007
In piers and chimney breasts, thickness					
75mm	m2	LD	0.75	9.90	0.005
100mm	m2	LD	0.80	9.90	0.006
140mm	m2	LD	0.90	9.90	0.007
215mm	m2	LD	1.00	9.90	0.007
In isolated casings, thickness					
75mm	m2	LD	0.95	9.90	0.005
100mm	m2	LD	1.00	9.90	0.006
140mm	m2	LD	1.10	9.90	0.007
215mm	m2	LD	1.20	9.90	0.007

	Unit	Labour grade	Labour hours	Blocks nr	Mortar m3
Extra for fair face and flush pointing					
one side	m2	LD	0.16	-	-
two sides	m2	LD	0.33	-	-
Bonding ends of blockwork to brickwork in alternate courses					
60mm	m2	LD	0.26	-	-
75mm	m2	LD	0.30	-	-
100mm	m2	LD	0.38	-	-
175mm	m2	LD	0.44	-	-
215mm	m2	LD	0.62	-	-

Precast concrete lightweight aggregate block in cement mortar (1:3)

In walls and partitions, thickness

	Unit	Labour grade	Labour hours	Blocks nr	Mortar m3
75mm	m2	LD	0.35	9.90	0.005
100mm	m2	LD	0.40	9.90	0.006
140mm	m2	LD	0.50	9.90	0.007
215mm	m2	LD	0.60	9.90	0.007

In skins of hollow walls, thickness

	Unit	Labour grade	Labour hours	Blocks nr	Mortar m3
75mm	m2	LD	0.45	9.90	0.005
100mm	m2	LD	0.50	9.90	0.006
140mm	m2	LD	0.60	9.90	0.007
215mm	m2	LD	0.70	9.90	0.007

	Unit	Labour grade	Labour hours	Blocks nr	Mortar m3
In piers and chimney breasts, thickness					
75mm	m2	LD	0.65	9.90	0.005
100mm	m2	LD	0.70	9.90	0.006
140mm	m2	LD	0.80	9.90	0.007
215mm	m2	LD	0.90	9.90	0.007
In isolated casings, thickness					
75mm	m2	LD	0.85	9.90	0.005
100mm	m2	LD	0.90	9.90	0.006
140mm	m2	LD	0.80	9.90	0.007
215mm	m2	LD	1.10	9.90	0.007
Extra for fair face and flush pointing					
one side	m2	LD	0.14	-	-
two sides	m2	LD	0.27	-	-
Bonding ends of blockwork to brickwork in alternate courses					
75mm	m2	LD	0.28	-	-
100mm	m2	LD	0.31	-	-
140mm	m2	LD	0.38	-	-
215mm	m2	LD	0.54	-	-

	Unit	Labour grade	Labour hours	Wall ties nr
Sundries				
Form 50mm cavity between skins of hollow walls with				
galvanised steel butterfly wall ties	m2	LD	0.10	3
galvanised steel twisted wall ties	m2	LD	0.10	3
stainless steel twisted wall ties	m2	LD	0.10	3
Form 75mm cavity between skins of hollow walls with				
galvanised steel butterfly wall ties	m2	LD	0.10	3
galvanised steel twisted wall ties	m2	LD	0.10	3
stainless steel twisted wall ties	m2	LD	0.10	3

	Unit	Labour grade	Labour hours
Hessian-based bitumen damp-proof course in gauged mortar (1:1:6)			
Horizontal width			
over 225mm	m2	LD	0.35
not exceeding 225mm	m2	LD	0.60

	Unit	Labour grade	Labour hours

Vertical width

over 225mm	m2	LD	0.40
not exceeding 225mm	m2	LD	0.70

Asbestos-based bitumen damp-proof course in gauged mortar (1:1:6)

Horizontal width

over 225mm	m2	LD	0.35
not exceeding 225mm	m2	LD	0.60

Vertical width

over 225mm	m2	LD	0.40
not exceeding 225mm	m2	LD	0.70

Pitch polymer damp-proof course in gauged mortar (1:1:6)

Horizontal width

over 225mm	m2	LD	0.35
not exceeding 225mm	m2	LD	0.60

Vertical width

over 225mm	m2	LD	0.40
not exceeding 225mm	m2	LD	0.70

	Unit	Labour grade	Labour hours
Asbestos-based with lead core damp-proof course in gauged mortar (1:1:6)			
Horizontal width			
over 225mm	m2	LD	0.40
not exceeding 225mm	m2	LD	0.70
Vertical width			
over 225mm	m2	LD	0.45
not exceeding 225mm	m2	LD	0.75
Slate damp-proof course bedded and pointed in 25mm cement mortar (1:3) over 225mm wide			
single course	m2	LA	0.20
double course	m2	LA	0.40
Slate damp-proof course bedded and pointed in 25mm cement mortar (1:3) not exceeding 225mm wide			
single course	m2	LA	0.45
double course	m2	LA	0.80

	Unit	Labour grade	Labour hours	DPC litre
Three coats bituminous waterproofing compound				
vertically	m2	LB	0.42	3.13
horizontally	m2	LB	0.33	3.13

	Unit	Labour grade	Labour hours
Cavity wall insulation sheets			
25mm	m2	LB	0.16
50mm	m2	LB	0.16

	Unit	Labour grade	Labour hours
Galvanised brick reinforcement, width			
65mm	m	LA	0.07
115mm	m	LA	0.09
175mm	m	LA	0.12
225mm	m	LA	0.15

	Unit	Labour grade	Labour hours	Mortar m3
Rake out joints of brickwork for flashing and point up on completion				
horizontal	m2	LA	0.09	0.0001
stepped	m2	LA	0.17	0.0002

	Unit	Labour grade	Labour hours

Chimney pots

Terracotta chimney pot, setting
and flaunching in cement mortar,
185mm diameter, height

300mm	nr	LA	1.30
375mm	nr	LA	1.50
450mm	nr	LA	1.80
600mm	nr	LA	2.00
750mm	nr	LA	2.40

Air bricks

Form opening in cavity wall
for air brick, seal cavity with
slates in cement mortar, size

225 x 75mm	nr	LA	0.30
225 x 150mm	nr	LA	0.42
225 x 225mm	nr	LA	0.54

Terracotta louvre pattern
air brick, size

215 x 65mm	nr	LA	0.10
215 x 140mm	nr	LA	0.10
215 x 215mm	nr	LA	0.10

Terracotta square hole
pattern air brick, size

215 x 65mm	nr	LA	0.10
215 x 140mm	nr	LA	0.10
215 x 215mm	nr	LA	0.10

	Unit	Labour grade	Labour hours
Cast iron louvre pattern air brick, size			
225 x 75mm	nr	LA	0.12
225 x 150mm	nr	LA	0.12
225 x 225mm	nr	LA	0.12

4

Stonework

Weights of materials	kg/m3
Cement	1400
Sand	1600
Lime, ground	750
Stone	
Artificial	2200
Bath	2200
Darley Dale	2400
Portland	2200
York	2400

Mortar per m2 of random rubble walling

Wall thickness	Mortar m3
300mm	0.120
450mm	0.160
550mm	0.200

Labour grades

2 Masons and 1 labourer	LE

	Unit	Labour grade	Labour hours	Stone tonne	Mortar m3

Random rubble walling, laid dry

	Unit	Labour grade	Labour hours	Stone tonne	Mortar m3
300mm	m2	LE	1.40	0.60	-
350mm	m2	LE	1.50	0.70	-
400mm	m2	LE	1.60	0.80	-
450mm	m2	LE	1.80	0.90	-
500mm	m2	LE	2.00	1.00	-

Random rubble walling in cement mortar (1:3)

	Unit	Labour grade	Labour hours	Stone tonne	Mortar m3
300mm	m2	LE	1.30	0.60	0.12
350mm	m2	LE	1.40	0.70	0.13
400mm	m2	LE	1.50	0.80	0.15
450mm	m2	LE	1.70	0.90	0.16
500mm	m2	LE	1.90	1.00	0.17

Irregular coursed rubble walling in cement mortar (1:3)

	Unit	Labour grade	Labour hours	Stone tonne	Mortar m3
300mm	m2	LE	1.40	0.60	0.12
350mm	m2	LE	1.50	0.70	0.13
400mm	m2	LE	1.60	0.80	0.15
450mm	m2	LE	1.80	0.90	0.16
500mm	m2	LE	2.00	1.00	0.17

Coursed rubble walling in cement mortar (1:3)

	Unit	Labour grade	Labour hours	Stone tonne	Mortar m3
300mm	m2	LE	1.30	0.60	0.12
350mm	m2	LE	1.40	0.70	0.13
400mm	m2	LE	1.50	0.80	0.15
450mm	m2	LE	1.70	0.90	0.16
500mm	m2	LE	1.90	1.00	0.17

	Unit	Labour grade	Labour hours	Stone tonne	Mortar m3
Fair return to walling, thickness					
300mm	m	LE	0.70	-	-
350mm	m	LE	0.75	-	-
400mm	m	LE	0.80	-	-
450mm	m	LE	0.90	-	-
500mm	m	LE	1.00	-	-
Prepare and level to receive damp-proof course on walling, thickness					
300mm	m	LE	0.30	-	-
350mm	m	LE	0.35	-	-
400mm	m	LE	0.40	-	-
450mm	m	LE	0.50	-	-
500mm	m	LE	0.60	-	-
Arch 250mm high, soffit width					
300mm	m	LE	1.50	-	-
350mm	m	LE	1.80	-	-
400mm	m	LE	2.20	-	-
450mm	m	LE	2.60	-	-
500mm	m	LE	3.00	-	-
Quoin stones, dressed on two faces, size					
250 x 200 x 300mm	m	LE	0.70	-	-
250 x 200 x 400mm	m	LE	0.80	-	-
250 x 250 x 300mm	m	LE	0.90	-	-
250 x 250 x 400mm	m	LE	1.10	-	-
350 x 200 x 300mm	m	LE	1.50	-	-
350 x 250 x 400mm	m	LE	1.80	-	-

	Unit	Labour grade	Labour hours	Stone tonne	Mortar m3
Natural dressed stone walling, bedded and jointed in gauged mortar, flush pointed one side, wall thickness					
50mm	m2	LE	2.80	0.11	0.004
75mm	m2	LE	3.00	0.17	0.005
100mm	m2	LE	3.20	0.22	0.006
Lintels, size					
200 x 100mm	m	LE	1.20	-	-
200 x 150mm	m	LE	1.40	-	-
Sills, sunk weathered and throated size					
200 x 75mm	m	LE	1.00	-	-
250 x 75mm	m	LE	1.10	-	-
300 x 75mm	m	LE	1.20	-	-
Band course, size					
175 x 175mm	m	LE	1.00	-	-
200 x 100mm	m	LE	1.10	-	-
Copings, twice weathered and twice throated size					
300 x 50mm	m	LE	0.70	-	-
300 x 75mm	m	LE	0.80	-	-
300 x 100mm	m	LE	0.90	-	-

	Unit	Labour grade	Labour hours	Stone tonne	Mortar m3
Reconstructed stone walling, bedded and jointed in gauged mortar, flush pointed one side, wall thickness					
50mm	m2	LE	2.80	0.11	0.004
75mm	m2	LE	3.00	0.17	0.005
100mm	m2	LE	3.20	0.22	0.006
Lintels, size					
200 x 100mm	m	LE	1.20	-	-
200 x 150mm	m	LE	1.40	-	-
Sills, sunk weathered and throated size					
200 x 75mm	m	LE	1.00	-	-
250 x 75mm	m	LE	1.10	-	-
300 x 75mm	m	LE	1.20	-	-
Band course, size					
175 x 175mm	m	LE	1.00	-	-
200 x 100mm	m	LE	1.10	-	-
Copings, twice weathered and twice throated size					
300 x 50mm	m	LE	0.70	-	-
300 x 75mm	m	LE	0.80	-	-
300 x 100mm	m	LE	0.90	-	-
Pier caps, size					
300 x 300mm	m	LE	0.30	-	-
400 x 300mm	m	LE	0.35	-	-

5

Asphalt work

Weights of materials	Thickness	Tonnes/m2
Cold asphalt		
one coat	13mm	0.03
	26mm	0.06
	50mm	0.12
	75mm	0.18
two coat	90mm	0.20
	100mm	0.23
Hot asphalt		
one coat	26mm	0.06
	50mm	0.12
	75mm	0.18
two coats	90mm	0.20
	100mm	0.24
Mastic asphalt	26mm	0.06
	50mm	0.12

Labour grades

2 Asphalt layers and 1 labourer   LF

Plant grades

Asphalt boiler   PG

	Unit	Labour grade	Labour hours	Plant grade	Plant hours	Materials tonne

Mastic asphalt tanking

13mm coat, laid horizontally

over 300mm wide	m2	LF	0.25	PG	0.15	0.030
not exceeding 150mm wide	m	LF	0.04	PG	0.02	0.005
150-300mm wide	m	LF	0.08	PG	0.04	0.010

26mm coat, laid horizontally

over 300mm wide	m2	LF	0.40	PG	0.25	0.060
not exceeding 150mm wide	m	LF	0.08	PG	0.04	0.010
150-300mm wide	m	LF	0.18	PG	0.08	0.020

13mm coat, laid vertically

over 300mm wide	m2	LF	0.35	PG	0.20	0.030
not exceeding 150mm wide	m	LF	0.05	PG	0.03	0.005
150-300mm wide	m	LF	0.09	PG	0.05	0.010

26mm coat, laid vertically

over 300mm wide	m2	LF	0.50	PG	0.30	0.060
not exceeding 150mm wide	m	LF	0.08	PG	0.05	0.010
150-300mm wide	m	LF	0.15	PG	0.10	0.020

	Unit	Labour grade	Labour hours	Plant grade	Plant hours	Materials tonne

Mastic asphalt flooring

15mm one coat light duty flooring

	Unit	Labour grade	Labour hours	Plant grade	Plant hours	Materials tonne
over 300mm wide	m2	LF	0.28	PG	0.16	0.030
not exceeding 150mm wide	m	LF	0.04	PG	0.03	0.005
150-300mm wide	m	LF	0.09	PG	0.05	0.010

20mm one coat medium duty flooring

	Unit	Labour grade	Labour hours	Plant grade	Plant hours	Materials tonne
over 300mm wide	m2	LF	0.32	PG	0.18	0.045
not exceeding 150mm wide	m	LF	0.05	PG	0.03	0.007
150-300mm wide	m	LF	0.11	PG	0.06	0.013

30mm one coat heavy duty flooring

	Unit	Labour grade	Labour hours	Plant grade	Plant hours	Materials tonne
over 300mm wide	m2	LF	0.34	PG	0.20	0.070
not exceeding 150mm wide	m	LF	0.06	PG	0.04	0.011
150-300mm wide	m	LF	0.12	PG	0.07	0.022

Mastic roofing

20mm two coat asphalt roofing

	Unit	Labour grade	Labour hours	Plant grade	Plant hours	Materials tonne
over 300mm wide	m2	LF	0.38	PG	0.22	0.045
not exceeding 150mm wide	m	LF	0.08	PG	0.05	0.007
150-300mm wide	m	LF	0.14	PG	0.08	0.013

6

Roofing and cladding

Weights of materials	kg/m2
Lead sheeting, 2.24mm thick	25.40
Aluminium, 0.80mm thick	2.20
Copper, 0.55mm thick	5.00
Zinc, 0.65mm thick	4.60

Profiled steel roof cladding (wide rib)

0.55 thick	5.30kg/m2
0.70 thick	6.74kg/m2
0.90 thick	8.67kg/m2

Roofing felt			kg/10m2
Type	3B	Glass fibre	18
	3E	Glass fibre	28
	3G	Glass fibre perforated underlay	32
Type	5B	Polyester base, mineral surfaced	38
	5E	Polyester base, mineral surfaced	38

	Lap mm	Gauge mm	nr/m2	Battens m/m2
Clay/concrete tiles				
267 x 165mm	65	100	60.00	10.00
	65	98	64.00	10.50
	65	90	68.00	11.30

	Lap mm	Gauge mm	nr/m2	Battens m/m2
Clay/concrete tiles				
387 x 230mm	75	300	16.00	3.20
	100	280	17.40	3.50
420 x 330mm	75	340	10.00	2.90
	100	320	10.74	3.10
Fibre slates				
500 x 250mm	90	205	19.50	4.85
	80	210	19.10	4.76
	70	215	18.60	4.65
600 x 300mm	105	250	13.60	4.04
	100	250	13.40	4.00
	90	255	13.10	3.92
	80	260	12.90	3.85
	70	263	12.70	3.77
400 x 200mm	70	165	30.00	6.06
	75	162	30.90	6.17
	90	155	32.30	6.45
500 x 250mm	70	215	18.60	4.65
	75	212	18.90	4.72
	90	205	19.50	4.88
	100	200	20.00	5.00
	110	195	20.50	5.13
600 x 300mm	100	250	13.40	4.00
	110	245	13.60	4.08
Natural slates				
405 x 205mm	75	165	29.59	8.70
405 x 255mm	75	165	23.75	6.06
405 x 305mm	75	165	19.00	5.00

	Lap mm	Gauge mm	nr/m2	Battens m/m2
460 x 230mm	75	195	23.00	6.00
460 x 255mm	75	195	20.37	5.20
460 x 305mm	75	195	17.00	5.00
510 x 255mm	75	220	18.02	4.60
510 x 305mm	75	220	15.00	4.00
560 x 280mm	75	240	14.81	4.12
560 x 305mm	75	240	14.00	4.00
610 x 305mm	75	265	12.27	3.74

Reconstructed stone slates

	Lap mm	Gauge mm	nr/m2	Battens m/m2
380 x 2150mm	75	150	16.00	3.20
	100	140	17.40	3.50

Steel round lost head nails per kg

75 x 3.75mm	160
65 x 3.35mm	240
65 x 3.00mm	270
60 x 3.35mm	270
60 x 3.00mm	330
50 x 3.00mm	360
50 x 2.65mm	420
40 x 2.36mm	760

Steel clout nails per kg

100 x 4.50mm	75
90 x 4.50mm	85
75 x 3.75mm	150
65 x 3.75mm	180
50 x 3.75mm	230
50 x 3.35mm	290
50 x 3.00mm	340
50 x 2.65mm	430
45 x 3.35mm	330

Steel clout nails per kg

45 x 2.65mm	440
40 x 3.35mm	350
40 x 2.65mm	570
40 x 2.36mm	700
30 x 3.00mm	540
30 x 2.65mm	660
30 x 2.36mm	830

Lead work	**Code**	**Colour**	**Thickness mm**	**kg/m2**
1.32mm	3	Green	1.32	14.97
1.80mm	4	Blue	1.80	20.41
2.24mm	5	Red	2.24	25.40
2.65mm	6	Black	2.65	30.05
3.15mm	7	White	3.15	35.72
3.55mm	8	Orange	3.55	40.26

Aluminium	**SWG**	**kg/m2**
0.60mm	23	1.54
0.80mm	21	2.05

Copper	**SWG**	**Sheet size mm**	**kg/m2**
0.45mm	26	600 x 1800	4.04
0.55mm	24	600 x 1800	4.94
0.70mm	22	750 x 1800	6.29

Zinc	**Gauge**	**kg/m2**
0.43mm	9	3.10
0.48mm	10	3.20
0.56mm	11	3.80
0.64mm	12	4.30
0.71mm	13	4.80
0.79mm	14	5.30
0.91mm	15	6.20
1.04mm	16	7.00

Roofing felt

			Code	kg/10m2
Type	1B	Fibre	White	18
	1E	Fibre	White	38
Type	2B	Asbestos fine gravel	Green	18
	2E	Asbestos mineral gravel	Green	38
Type	3B	Glass fibre	Red	18
	3E	Glass fibre	Red	28
	36	Glass fibre perforated underlay	Red	32

Labour grades

Semi-skilled operative   LB

1 Roofer and 1 labourer   LG

	Unit	Labour grade	Labour hours	Tiles nr	Battens m
Clay/concrete tiling					
Roof tiles size 267 x 165mm, 65mm lap, 35 degrees pitch					
Battens size 38 x 19mm					
gauge 100mm	m2	LG	1.00	60.00	10.00
gauge 95mm	m2	LG	1.02	64.00	10.00
gauge 90mm	m2	LG	1.04	68.00	11.30
Battens size 38 x 25mm					
gauge 100mm	m2	LG	1.10	60.00	10.00
gauge 95mm	m2	LG	1.12	64.00	10.00
Extra for					
nailing every tile	m2	LG	0.30	-	-
double eaves course	m	LG	0.35	-	-
ridge tile	m	LG	0.35	-	-
double eaves course	m	LG	0.40	-	-
hip tile	m	LG	0.60	-	-
vent terminal	m	LG	0.60	-	-
straight cutting	m	LG	0.20	-	-
forming hole for pipe	nr	LG	0.40	-	-
Roof tiles size 387 x 229mm, battens size 38 x 25mm					
75mm lap, pitch 25 to 44 degrees	m2	LG	0.60	16.00	3.20
100mm lap, pitch 22 to 44 degrees	m2	LG	0.60	16.00	3.20

	Unit	Labour grade	Labour hours	Tiles nr	Battens m
Extra for					
nailing every tile	m2	LG	0.30	-	-
double eaves course	m	LG	0.35	-	-
ridge tile	m	LG	0.35	-	-
double eaves course	m	LG	0.40	-	-
hip tile	m	LG	0.60	-	-
vent terminal	m	LG	0.60	-	-
straight cutting	m	LG	0.20	-	-
forming hole for pipe	nr	LG	0.40	-	-

Roof tiles size 420 x 330mm, battens size 38 x 25mm

	Unit	Labour grade	Labour hours	Tiles nr	Battens m
75mm lap, pitch 25 to 44 degrees	m2	LG	0.56	10.00	2.90
100mm lap, pitch 22.5 to 44 degrees	m2	LG	0.58	10.74	3.10
Extra for					
nailing every tile	m2	LG	0.05	-	-
double eaves course	m	LG	0.35	-	-
ridge tile	m	LG	0.35	-	-
double eaves course	m	LG	0.40	-	-
hip tile	m	LG	0.60	-	-
vent terminal	m	LG	0.60	-	-
straight cutting	m	LG	0.20	-	-
forming hole for pipe	nr	LG	0.40	-	-

	Unit	Labour grade	Labour hours	Tiles nr	Battens m

Fibre cement slating

Non-asbestos fibre
cement slates size 500
x 250mm, pitch over
40 degrees, 38 x 25mm
softwood battens, type
1F reinforced underlay

lap 70mm, gauge 215mm	m2	LG	0.65	18.60	4.65

Non-asbestos fibre
cement slates size 500
x 250mm, pitch over
27.5 degrees, 38 x 25mm
softwood battens, type
1F reinforced underlay

lap 80mm, gauge 210mm	m2	LG	0.66	19.10	4.76

Non-asbestos fibre
cement slates size 500
x 250mm, pitch over
25 degrees, 38 x 25mm
softwood battens, type
1F reinforced underlay

lap 90mm, gauge 205mm	m2	LG	0.68	19.50	4.48

Non-asbestos fibre
cement slates size 600
x 300mm, pitch over
35 degrees, 38 x 25mm
softwood battens, type
1F reinforced underlay

lap 70mm, gauge 265mm	m2	LG	0.55	12.70	3.77

	Unit	Labour grade	Labour hours	Tiles nr	Battens m
Non-asbestos fibre cement slates size 600 x 300mm, pitch over 30 degrees, 38 x 25mm softwood battens, type 1F reinforced underlay					
lap 100mm, gauge 250mm	m2	LG	0.55	13.40	4.00
Non-asbestos fibre cement slates size 600 x 300mm, pitch 25 degrees, 38 x 25mm softwood battens, type 1F reinforced underlay					
lap 110mm, gauge 245mm	m2	LG	0.55	13.60	4.08
Non-asbestos fibre cement slates size 400 x 200mm, pitch over 40 degrees, 38 x 25mm softwood battens, type 1F reinforced underlay					
lap 90mm, gauge 155mm	m2	LG	0. 85	32.30	6.45
Non-asbestos fibre cement slates size 400 x 200mm, pitch 40 to 45 degrees, 38 x 25mm softwood battens, type 1F reinforced underlay					
lap 70mm, gauge 155mm	m2	LG	0.88	30.00	6.06

	Unit	Labour grade	Labour hours	Tiles nr	Battens m
Natural slating					
Blue/grey slates size 405 x 205mm, 75mm lap, 50 x 25mm softwood battens, type 1F reinforced underlay					
Sloping	m2	LG	1.00	29.59	8.70
Vertical	m2	LG	1.10	29.59	8.70
Mansard	m2	LG	1.10	29.59	8.70
Extra for					
ridge tile	m	LG	0.70	-	-
double eaves course	m	LG	0.50	-	-
hip tile	m	LG	0.70	-	-
straight cutting	m	LG	0.60	-	-
forming hole for pipe	nr	LG	0.40	-	-
Blue/grey slates size 405 x 255mm, 75mm lap, 50 x 25mm softwood battens, type 1F reinforced underlay					
Sloping	m2	LG	0.80	23.75	6.06
Vertical	m2	LG	0.90	23.75	6.06
Mansard	m2	LG	0.90	23.75	6.06
Extra for					
ridge tile	m	LG	0.70	-	-
double eaves course	m	LG	0.50	-	-
hip tile	m	LG	0.70	-	-
straight cutting	m	LG	0.60	-	-
forming hole for pipe	nr	LG	0.40	-	-

	Unit	Labour grade	Labour hours	Tiles nr	Battens m
Blue/grey slates size 405 x 305mm, 75mm lap, 50 x 25mm softwood battens, type 1F reinforced underlay					
Sloping	m2	LG	0.75	19.00	5.00
Vertical	m2	LG	0.85	19.00	5.00
Mansard	m2	LG	0.85	19.00	5.00
Extra for					
ridge tile	m	LG	0.70	-	-
double eaves course	m	LG	0.50	-	-
hip tile	m	LG	0.70	-	-
straight cutting	m	LG	0.60	-	-
forming hole for pipe	nr	LG	0.40	-	-
Blue/grey slates size 460 x 230mm, 75mm lap, 50 x 25mm softwood battens, type 1F reinforced underlay					
Sloping	m2	LG	0.80	23.06	6.00
Vertical	m2	LG	0.90	23.06	6.00
Mansard	m2	LG	0.90	23.06	6.00
Extra for					
ridge tile	m	LG	0.70	-	-
double eaves course	m	LG	0.50	-	-
hip tile	m	LG	0.70	-	-
straight cutting	m	LG	0.60	-	-
forming hole for pipe	nr	LG	0.40	-	-

	Unit	Labour grade	Labour hours	Tiles nr	Battens m
Blue/grey slates size 460 x 255mm, 75mm lap, 50 x 25mm softwood battens, type 1F reinforced underlay					
Sloping	m2	LG	0.75	20.37	5.20
Vertical	m2	LG	0.85	20.37	5.20
Mansard	m2	LG	0.85	20.37	5.20
Extra for					
ridge tile	m	LG	0.70	-	-
double eaves course	m	LG	0.50	-	-
hip tile	m	LG	0.70	-	-
straight cutting	m	LG	0.60	-	-
forming hole for pipe	nr	LG	0.40	-	-
Blue/grey slates size 460 x 305mm, 75mm lap, 50 x 25mm softwood battens, type 1F reinforced underlay					
Sloping	m2	LG	0.70	17.00	5.00
Vertical	m2	LG	0.80	17.00	5.00
Mansard	m2	LG	0.80	17.00	5.00
Extra for					
ridge tile	m	LG	0.70	-	-
double eaves course	m	LG	0.50	-	-
hip tile	m	LG	0.70	-	-
straight cutting	m	LG	0.60	-	-
forming hole for pipe	nr	LG	0.40	-	-

	Unit	Labour grade	Labour hours	Tiles nr	Battens m
Blue/grey slates size 510 x 255mm, 75mm lap, 50 x 25mm softwood battens, type 1F reinforced underlay					
Sloping	m2	LG	0.73	18.02	4.60
Vertical	m2	LG	0.85	18.02	4.60
Mansard	m2	LG	0.85	18.02	4.60
Extra for					
ridge tile	m	LG	0.70	-	-
double eaves course	m	LG	0.50	-	-
hip tile	m	LG	0.70	-	-
straight cutting	m	LG	0.60	-	-
forming hole for pipe	nr	LG	0.40	-	-
Blue/grey slates size 510 x 305mm, 75mm lap, 50 x 25mm softwood battens, type 1F reinforced underlay					
Sloping	m2	LG	0.60	15.00	4.00
Vertical	m2	LG	0.70	15.00	4.00
Mansard	m2	LG	0.70	15.00	4.00
Extra for					
ridge tile	m	LG	0.70	-	-
double eaves course	m	LG	0.50	-	-
hip tile	m	LG	0.70	-	-
straight cutting	m	LG	0.60	-	-
forming hole for pipe	nr	LG	0.40	-	-

	Unit	Labour grade	Labour hours	Tiles nr	Battens m
Blue/grey slates size 560 x 280mm, 75mm lap, 50 x 25mm softwood battens, type 1F reinforced underlay					
Sloping	m2	LG	0.60	14.81	4.12
Extra for					
ridge tile	m	LG	0.70	-	-
double eaves course	m	LG	0.50	-	-
hip tile	m	LG	0.70	-	-
straight cutting	m	LG	0.60	-	-
forming hole for pipe	nr	LG	0.40	-	-
Blue/grey slates size 560 x 305mm, 75mm lap, 50 x 25mm softwood battens, type 1F reinforced underlay					
Sloping	m2	LG	0.55	14.00	4.00
Extra for					
ridge tile	m	LG	0.70	-	-
double eaves course	m	LG	0.50	-	-
hip tile	m	LG	0.70	-	-
straight cutting	m	LG	0.60	-	-
forming hole for pipe	nr	LG	0.40	-	-

	Unit	Labour grade	Labour hours	Tiles nr	Battens m
Reconstructed stone slating					
Interlocking slate 380 x 250mm type 1F reinforced underlay					
75mm lap, pitch 25 to 90 degrees	m2	LG	0.75	16.00	3.20
100mm lap, pitch 25 to 90 degrees	m2	LG	0.78	17.40	3.50

	Unit	Labour grade	Labour hours	Thickness mm	Materials kg/m2
Lead sheet coverings					
Roof coverings, milled sheet lead					
Flat roofing, pitch less than 10 degrees to the horizontal					
code 4	m2	LG	4.00	1.80	20.41
code 5	m2	LG	4.20	2.24	25.40
code 6	m2	LG	4.40	2.65	30.05
code 7	m2	LG	4.60	3.15	35.72
code 8	m2	LG	4.80	3.55	40.26
Flashings, code 4, girth					
150mm	m	LG	0.45	1.80	3.06
200mm	m	LG	0.50	1.80	4.08
300mm	m	LG	0.55	1.80	6.12

	Unit	Labour grade	Labour hours	Thickness mm	Materials kg/m2
Aprons and sills, code 4, girth					
200mm	m	LG	0.50	1.80	4.08
300mm	m	LG	0.75	1.80	6.12
400mm	m	LG	1.00	1.80	8.16
Valleys and gutters, code 4, girth					
400mm	m	LG	1.00	1.80	8.16
600mm	m	LG	1.20	1.80	12.25
800mm	m	LG	1.40	1.80	16.33
Slates, size 400 x 400mm with collar 200mm high x 100mm diameter, code 4	nr	LG	1.50	1.80	3.85
Flashings, code 5, girth					
150mm	m	LG	0.45	2.24	3.81
200mm	m	LG	0.50	2.24	5.08
300mm	m	LG	0.55	2.24	7.62
Aprons and sills, code 5, girth					
200mm	m	LG	0.50	2.24	5.08
300mm	m	LG	0.75	2.24	7.62
400mm	m	LG	1.00	2.24	10.16
Valleys and gutters, code 5, girth					
400mm	m	LG	1.00	2.24	10.16
600mm	m	LG	1.20	2.24	15.24
800mm	m	LG	1.40	2.24	20.32

	Unit	Labour grade	Labour hours	Thickness mm	Materials kg/m2
Slates, size 400 x 400mm with collar 200mm high x 100mm diameter, code 5	nr	LG	1.50	2.24	4.85

Aluminium sheet coverings

0.60mm commercial grade aluminium sheeting in roof

	Unit	Labour grade	Labour hours	Thickness mm	Materials kg/m2
flat	m2	LG	2.90	0.61	1.54
sloping 10 degrees to 50 degrees	m2	LG	3.20	0.61	1.54
sloping or vertical over 50 degrees	m2	LG	3.50	0.61	1.54

Flashings, girth

	Unit	Labour grade	Labour hours	Thickness mm	Materials kg/m2
150mm	m	LG	0.50	0.61	0.23
200mm	m	LG	0.55	0.61	0.31
300mm	m	LG	0.60	0.61	0.46

Aprons and sills, girth

	Unit	Labour grade	Labour hours	Thickness mm	Materials kg/m2
200mm	m	LG	0.60	0.61	0.31
300mm	m	LG	0.65	0.61	0.46
400mm	m	LG	0.90	0.61	0.62

Valleys and gutters, girth

	Unit	Labour grade	Labour hours	Thickness mm	Materials kg/m2
400mm	m	LG	1.00	2.65	12.02
600mm	m	LG	1.15	2.65	18.03
800mm	m	LG	1.30	2.65	24.04

	Unit	Labour grade	Labour hours	Thickness mm	Materials kg/m2
0.80mm commercial grade aluminium sheeting in roof					
flat	m2	LG	3.00	0.80	2.05
sloping 10 degrees to 50 degrees	m2	LG	3.33	0.80	2.05
sloping or vertical over 50 degrees	m2	LG	3.60	0.80	2.05
Flashings, girth					
150mm	m	LG	0.50	0.80	0.31
200mm	m	LG	0.55	0.80	0.41
300mm	m	LG	0.60	0.80	0.62
Aprons and sills, girth					
200mm	m	LG	0.60	0.80	0.41
300mm	m	LG	0.65	0.80	0.62
400mm	m	LG	0.90	0.80	0.82
Valleys and gutters, girth					
400mm	m	LG	1.00	0.80	0.82
600mm	m	LG	1.15	0.80	1.23
800mm	m	LG	1.30	0.80	1.64

Copper sheet coverings

0.45mm thick copper sheeting in roof					
flat	m2	LG	3.10	0.45	4.04
sloping 10 degrees to 50 degrees	m2	LG	3.33	0.45	4.04
sloping or vertical over 50 degrees	m2	LG	3.60	0.45	4.04

	Unit	Labour grade	Labour hours	Thickness mm	Materials kg/m2
Flashings, girth					
150mm	m	LG	0.50	0.45	0.61
200mm	m	LG	0.55	0.45	0.81
300mm	m	LG	0.60	0.45	1.21
Aprons and sills, girth					
200mm	m	LG	0.60	0.45	0.81
300mm	m	LG	0.65	0.45	1.21
400mm	m	LG	0.90	0.45	1.62
Valleys and gutters, girth					
400mm	m	LG	1.00	0.80	1.62
600mm	m	LG	1.15	0.80	2.42
800mm	m	LG	1.30	0.80	3.23
0.55mm thick copper sheeting in roof					
flat	m2	LG	3.20	0.55	4.94
sloping 10 degrees to 50 degrees	m2	LG	3.40	0.55	4.94
sloping or vertical over 50 degrees	m2	LG	3.90	0.55	4.94
Flashings, girth					
150mm	m	LG	0.50	0.55	0.74
200mm	m	LG	0.55	0.55	0.99
300mm	m	LG	0.60	0.55	1.48
Aprons and sills, girth					
200mm	m	LG	0.60	0.55	0.99
300mm	m	LG	0.65	0.55	1.48
400mm	m	LG	0.90	0.55	1.98

	Unit	Labour grade	Labour hours	Thickness mm	Materials kg/m2
Valleys and gutters, girth					
400mm	m	LG	1.00	0.55	1.98
600mm	m	LG	1.15	0.55	2.96
800mm	m	LG	1.30	0.55	3.95
0.70mm thick copper sheeting in roof					
flat	m2	LG	3.20	0.70	6.29
sloping 10 degrees to 50 degrees	m2	LG	3.60	0.70	6.29
sloping or vertical over 50 degrees	m2	LG	4.10	0.70	6.29
Flashings, girth					
150mm	m	LG	0.50	0.70	0.94
200mm	m	LG	0.55	0.70	1.26
300mm	m	LG	0.60	0.70	1.89
Aprons and sills, girth					
200mm	m	LG	0.60	0.70	1.26
300mm	m	LG	0.65	0.70	1.89
400mm	m	LG	0.90	0.70	2.52
Valleys and gutters, girth					
400mm	m	LG	1.00	0.70	2.52
600mm	m	LG	1.15	0.70	3.77
800mm	m	LG	1.30	0.70	5.03

	Unit	Labour grade	Labour hours	Thickness mm	Materials kg/m2
Zinc sheet coverings					
0.64mm thick zinc (grade 12) sheeting in roof					
flat	m2	LG	3.00	0.64	4.30
sloping 10 degrees to 50 degrees	m2	LG	3.30	0.64	4.30
sloping or vertical over 50 degrees	m2	LG	3.80	0.64	4.30
Flashings, girth					
150mm	m	LG	0.40	0.64	0.65
200mm	m	LG	0.45	0.64	0.86
300mm	m	LG	0.50	0.64	1.29
Aprons and sills, girth					
200mm	m	LG	0.50	0.64	0.86
300mm	m	LG	0.55	0.64	1.29
400mm	m	LG	0.80	0.64	1.72
Valleys and gutters, girth					
400mm	m	LG	0.90	0.64	1.72
600mm	m	LG	1.05	0.64	2.58
800mm	m	LG	1.20	0.64	3.44
0.70mm thick zinc sheeting in roof					
flat	m2	LG	3.20	0.70	6.29
sloping 10 degrees to 50 degrees	m2	LG	3.50	0.64	4.30
sloping or vertical over 50 degrees	m2	LG	4.00	0.64	4.30

	Unit	Labour grade	Labour hours	Thickness mm	Materials kg/m2
Flashings, girth					
150mm	m	LG	0.40	0.70	0.94
200mm	m	LG	0.45	0.70	1.25
300mm	m	LG	0.50	0.70	1.89
Aprons and sills, girth					
200mm	m	LG	0.50	0.70	1.25
300mm	m	LG	0.55	0.70	1.89
400mm	m	LG	0.80	0.70	2.52
Valleys and gutters, girth					
400mm	m	LG	0.90	0.70	2.52
600mm	m	LG	1.05	0.70	3.77
800mm	m	LG	1.20	0.70	5.03
0.79mm thick zinc (grade 14) sheeting in roof					
flat	m2	LG	3.10	0.79	5.30
sloping 10 degrees to 50 degrees	m2	LG	3.40	0.79	5.30
sloping or vertical over 50 degrees	m2	LG	3.90	0.79	5.30
Flashings, girth					
150mm	m	LG	0.40	0.79	0.80
200mm	m	LG	0.45	0.79	1.06
300mm	m	LG	0.50	0.79	1.59
Aprons and sills, girth					
200mm	m	LG	0.50	0.79	1.06
300mm	m	LG	0.55	0.79	1.59
400mm	m	LG	0.80	0.79	2.12

	Unit	Labour grade	Labour hours	Thickness mm	Materials kg/m2
Valleys and gutters, girth					
400mm	m	LG	0.90	0.79	2.12
600mm	m	LG	1.05	0.79	3.18
800mm	m	LG	1.20	0.79	4.24

	Unit	Labour grade	Labour hours	Materials kg/m2
Built-up felt roof coverings				
Built-up bituminous felt roofing coverings, layers fully bonded with hot bitumen laid to 5 degrees pitch				
Fibre-based sand surfaced felt type 1B (14kg/10m2)				
one layer	m2	LG	0.22	1.40
Fibre-based sand surfaced felt type 1B (18kg/10m2)				
one layer	m2	LG	0.23	1.80
two layers	m2	LG	0.30	3.60
three layers	m2	LG	0.45	5.40
Fibre-based sand surfaced felt type 1B (25kg/10m2)				
one layer	m2	LG	0.24	2.50
two layers	m2	LG	0.32	5.00
three layers	m2	LG	0.47	7.50

	Unit	Labour grade	Labour hours	Materials kg/m2
Fibre-based mineral surfaced felt type 1E (38kg/10m2)				
one layer	m2	LG	0.23	3.80
Glass fibre-based sand surfaced felt type 3B (28kg/10m2)				
one layer	m2	LG	0.23	1.80
two layers	m2	LG	0.30	3.60
three layers	m2	LG	0.45	5.40
Fibre-based mineral surfaced felt type 1E				
one layer	m2	LG	0.26	2.80
Glass-fibre based venting layer felt type 3G (28kg/10m2)				
one layer	m2	LG	0.28	3.20
Polyester based sand surfaced felt type 5V (29kg/10m2)				
one layer	m2	LG	0.25	2.90
Polyester-based sand surfaced felt type 5B (34kg/10m2)				
one layer	m2	LG	0.26	3.40

	Unit	Labour grade	Labour hours	Materials kg/m2
Polyester-based mineral surfaced felt type 5E (38kg/10m2)				
one layer	m2	LG	0.28	3.80
Polyester-based sand surfaced elastomeric bitumen coated felt (40kg/10m2)				
one layer	m2	LG	0.24	4.00
Polyester-based mineral surfaced elastomeric bitumen coated felt (32kg/10m2)				
one layer	m2	LG	0.28	3.20

Rooflights

Single skin glazed
PVC-U rooflight
domed plugged and
screwed to kerb,
diameter

600mm	nr	LB	0.60	-
900mm	nr	LB	0.70	-
1200mm	nr	LB	0.80	-
1800mm	nr	LB	0.90	-
600 x 600mm	nr	LB	1.00	-
780 x 780mm	nr	LB	1.10	-
900 x 900mm	nr	LB	1.25	-
1000 x 1000mm	nr	LB	1.50	-
1250 x 1250mm	nr	LB	1.50	-
1800 x 1800mm	nr	LB	1.50	-

	Unit	Labour grade	Labour hours	Materials kg/m2

Cladding

	Unit	Labour grade	Labour hours	Materials kg/m2
Galvanised steel troughed sheeting 0.7mm thick, fixed vertically to steel rails	m2	LG	0.50	-
Galvanised steel profiled sheeting 0.7mm thick, fixed vertically to steel rails	m2	LG	0.55	-
Aluminium troughed sheeting 0.7mm thick, fixed vertically to steel rails	m2	LG	0.55	-
PVC-U cladding in shiplap planks 100mm wide	m2	LG	0.75	-
PVC-U cladding in shiplap planks 150mm wide	m2	LG	0.65	-

7

Carpentry and joinery

Weights of materials	kg/m3
Blockboard	
standard	940-1000
tempered	940-1060
Wood chipboard	
standard grade	650-750
flooring grade	680-800
Laminboard	500-700
Timber	
Ash	800
Baltic Spruce	480
Beech	816
Birch	720
Box	961
Cedar	480
Ebony	1217
Elm	624
Greenheart	961
Jarrah	816
Maple	752
Oak, American	720
Oak, English	848
Pine, Pitchpine	800
Pine, Red Deal	576
Pine, Yellow Deal	528

Weights of materials   kg/m3

Sycamore	530
Teak, African	961
Teak, Indian	656
Walnut	496

Number of nails per kg   nr

Oval brad or lost head nails

150 x 7.10 x 5.00	31
125 x 6.70 x 4.50	44
100 x 6.00 x 4.00	64
75 x 5.00 x 3.35	125
65 x 4.00 x 2.65	230
60 x 3.75 x 2.36	340
50 x 3.35 x 2.00	470
40 x 2.65 x 1.60	940
30 x 2.65 x 1.60	1480
25 x 2.00 x 1.25	2530

Round plain head nails

150 x 6.00	29
125 x 5.60	42
125 x 5.00	53
115 x 5.00	57
100 x 5.00	66
100 x 4.50	77
100 x 4.00	88
90 x 4.00	106
75 x 4.00	121
75 x 3.75	154
75 x 3.35	194
65 x 3.35	230
65 x 3.00	275
65 x 2.65	350
60 x 3.35	255

Number of nails per kg   **nr**

60 x 3.00	310
60 x 2.65	385
50 x 3.35	290
50 x 3.00	340
50 x 2.65	440
50 x 2.36	550

Round plain head nails

45 x 2.65	510
45 x 2.36	640
40 x 2.65	575
40 x 2.36	750
40 x 2.00	970
30 x 2.36	840
30 x 2.00	1170
25 x 2.00	1430
25 x 1.80	1720
25 x 1.60	2210
20 x 1.60	2710

Round lost head nails

65 x 3.35	240
65 x 3.00	270
75 x 3.75	160
60 x 3.35	270
60 x 3.00	330
50 x 3.00	360
40 x 2.36	760

Panel pins

50 x 2.00	770
40 x 1.60	1590
30 x 1.60	1900
25 x 1.60	2340

Number of nails per kg

	nr
25 x 1.40	3090
20 x 1.60	3140
20 x 1.40	3970

Lengths of boarding required

Effective width mm	m/m2
75	13.33
100	10.00
125	8.00
150	6.67
175	5.71
200	5.00

Labour grades

Craftsman	LA

	Unit	Labour grade	Labour hours	Materials m3	Nails kg

Carpentry

Sawn softwood, untreated

Floors

50 x 100mm	m	LA	0.22	0.0050	0.005
50 x 125mm	m	LA	0.25	0.0063	0.007
50 x 125mm	m	LA	0.28	0.0094	0.010
75 x 150mm	m	LA	0.30	0.0113	0.012

Partitions

38 x 75mm	m	LA	0.32	0.0029	0.025
38 x 100mm	m	LA	0.34	0.0038	0.033
50 x 75mm	m	LA	0.34	0.0038	0.033
50 x 100mm	m	LA	0.34	0.0150	0.044

Flat roofs

38 x 100mm	m	LA	0.15	0.0038	0.003
50 x 75mm	m	LA	0.17	0.0038	0.003
50 x 100mm	m	LA	0.18	0.0050	0.005
50 x 125mm	m	LA	0.19	0.0063	0.007
50 x 150mm	m	LA	0.20	0.0075	0.008
75 x 100mm	m	LA	0.20	0.0075	0.009
75 x 125mm	m	LA	0.26	0.0094	0.010

Pitched roofs

38 x 100mm	m	LA	0.22	0.0038	0.013
50 x 75mm	m	LA	0.24	0.0038	0.013
50 x 100mm	m	LA	0.25	0.0050	0.017
50 x 125mm	m	LA	0.26	0.0063	0.022
50 x 150mm	m	LA	0.27	0.0075	0.026
75 x 100mm	m	LA	0.27	0.0075	0.026
75 x 125mm	m	LA	0.38	0.0094	0.032

	Unit	Labour grade	Labour hours	Materials m3	Nails kg
Kerb, bearer					
25 x 75mm	m	LA	0.12	0.0019	0.008
25 x 100mm	m	LA	0.16	0.0025	0.009
25 x 150mm	m	LA	0.19	0.0038	0.010
38 x 75mm	m	LA	0.15	0.0028	0.011
38 x 100mm	m	LA	0.20	0.0038	0.012
50 x 50mm	m	LA	0.14	0.0025	0.012
50 x 75mm	m	LA	0.20	0.0038	0.013
50 x 100mm	m	LA	0.26	0.0050	0.017
75 x 75mm	m	LA	0.28	0.0056	0.020
75 x 100mm	m	LA	0.34	0.0075	0.026
75 x 125mm	m	LA	0.42	0.0094	0.032
Solid strutting					
38 x 100mm	m	LA	0.20	0.0038	0.030
50 x 100mm	m	LA	0.14	0.0050	0.060
50 x 125mm	m	LA	0.20	0.0063	0.070
50 x 150mm	m	LA	0.26	0.0075	0.090
Herringbone strutting **50 x 50mm to joists, depth**					
125mm	m	LA	0.60	0.0042	0.055
150mm	m	LA	0.60	0.0047	0.060
175mm	m	LA	0.60	0.0050	0.065
240mm	m	LA	0.60	0.0056	0.070
Trimming around rectangular openings, joists size					
50 x 100mm	m	LA	1.50	-	-
50 x 125mm	m	LA	1.65	-	-
50 x 150mm	m	LA	1.80	-	-
75 x 125mm	m	LA	2.12	-	-
75 x 150mm	m	LA	2.35	-	-

	Unit	Labour grade	Labour hours	Materials m2	Nails kg
Plywood marine quality in gutters, eaves, verges, soffits and fascias, 12mm thickness, width					
Over 300mm	m2	LA	1.30	1.00	0.060
150mm	m	LA	0.30	0.15	0.012
175mm	m	LA	0.33	0.18	0.013
200mm	m	LA	0.35	0.20	0.014
225mm	m	LA	0.37	0.23	0.015
250mm	m	LA	0.40	0.25	0.016
Plywood marine quality in gutters, eaves, verges, soffits and fascias, 18mm thickness, width					
Over 300mm	m2	LA	1.60	1.00	0.060
150mm	m	LA	0.35	0.15	0.012
175mm	m	LA	0.37	0.18	0.013
200mm	m	LA	0.40	0.20	0.014
225mm	m	LA	0.43	0.23	0.015
250mm	m	LA	0.45	0.25	0.016
Plywood marine quality in gutters, eaves, verges, soffits and fascias, 25mm thickness, width					
Over 300mm	m2	LA	1.80	1.00	0.060
150mm	m	LA	0.40	0.15	0.012
175mm	m	LA	0.42	0.18	0.013
200mm	m	LA	0.45	0.20	0.014
225mm	m	LA	0.48	0.23	0.015
250mm	m	LA	0.50	0.25	0.016

	Unit	Labour grade	Labour hours	Materials m2	Nails kg
Raking cutting on marine quality plywood, thickness					
12mm	m	LA	0.30	-	-
18mm	m	LA	0.35	-	-
25mm	m	LA	0.40	-	-
Curved cutting on marine quality plywood, thickness					
12mm	m	LA	0.50	-	-
18mm	m	LA	0.55	-	-
25mm	m	LA	0.60	-	-
Wrought softwood in gutters, eaves, verges, soffits and fascias, 13mm thickness, width					
Over 300mm	m2	LA	1.40	1.00	0.060
150mm	m	LA	0.35	0.15	0.012
175mm	m	LA	0.40	0.18	0.013
200mm	m	LA	0.45	0.20	0.014
225mm	m	LA	0.50	0.23	0.015
250mm	m	LA	0.55	0.25	0.016
Wrought softwood in gutters, eaves, verges, soffits and fascias, 19mm thickness, width					
Over 300mm	m2	LA	1.70	1.00	0.060
150mm	m	LA	0.40	0.15	0.012
175mm	m	LA	0.45	0.18	0.013
200mm	m	LA	0.50	0.20	0.014
225mm	m	LA	0.55	0.23	0.015
250mm	m	LA	0.60	0.25	0.016

	Unit	Labour grade	Labour hours	Materials m2	Nails kg
Wrought softwood in gutters, eaves, verges, soffits and fascias, 25mm thickness, width					
Over 300mm	m2	LA	1.90	1.00	0.060
150mm	m	LA	0.50	0.15	0.012
175mm	m	LA	0.55	0.18	0.013
200mm	m	LA	0.60	0.20	0.014
225mm	m	LA	0.65	0.23	0.015
250mm	m	LA	0.70	0.25	0.016
Raking cutting on wrought softwood, thickness					
13mm	m	LA	0.30	-	-
19mm	m	LA	0.35	-	-
25mm	m	LA	0.40	-	-
Curved cutting on wrought softwood, thickness					
13mm	m	LA	0.50	-	-
19mm	m	LA	0.55	-	-
25mm	m	LA	0.60	-	-
Non-asbestos boarding in gutters, eaves, verges, soffits and fascias, 6mm thickness, width					
Over 300mm	m2	LA	1.50	1.00	0.060
150mm	m	LA	0.40	0.15	0.012
175mm	m	LA	0.45	0.18	0.013
200mm	m	LA	0.50	0.20	0.014
225mm	m	LA	0.55	0.23	0.015
250mm	m	LA	0.60	0.25	0.016

	Unit	Labour grade	Labour hours	Materials m2	Nails kg
Non-asbestos boarding in gutters, eaves, verges, soffits and fascias, 9mm thickness, width					
Over 300mm	m2	LA	1.75	1.00	0.060
150mm	m	LA	0.45	0.15	0.012
175mm	m	LA	0.50	0.18	0.013
200mm	m	LA	0.55	0.20	0.014
225mm	m	LA	0.50	0.23	0.015
250mm	m	LA	0.60	0.25	0.016
Non-asbestos boarding in gutters, eaves, verges, soffits and fascias, 12mm thickness, width					
Over 300mm	m2	LA	1.95	1.00	0.060
150mm	m	LA	0.55	0.15	0.012
175mm	m	LA	0.60	0.18	0.013
200mm	m	LA	0.65	0.20	0.014
225mm	m	LA	0.70	0.23	0.015
250mm	m	LA	0.75	0.25	0.016
Raking cutting on non-asbestos boarding, thickness					
6mm	m	LA	0.30	-	-
9mm	m	LA	0.35	-	-
12mm	m	LA	0.40	-	-
Curved cutting on non-asbestos boarding, thickness					
6mm	m	LA	0.50	-	-
9mm	m	LA	0.55	-	-
12mm	m	LA	0.60	-	-

	Unit	Labour grade	Labour hours	Materials m2	Nails kg

Trussed rafters

Gang-nailed trussed rafter
(Fink pattern), 22.5, 30 or
45 degrees pitch, 450mm
overhang, span

	Unit	Labour grade	Labour hours	Materials m2	Nails kg
5m	nr	LA	1.50	-	-
6m	nr	LA	1.60	-	-
7m	nr	LA	1.70	-	-
8m	nr	LA	1.80	-	-
9m	nr	LA	1.90	-	-
10m	nr	LA	2.00	-	-

Mono-pitch gang-nailed
trussed rafter, 17.5, 30 or
45 degrees pitch, 450mm
overhang, span

	Unit	Labour grade	Labour hours	Materials m2	Nails kg
3m	nr	LA	1.10	-	-
4m	nr	LA	1.20	-	-
5m	nr	LA	1.30	-	-
6m	nr	LA	1.40	-	-

Glued laminated beams,
size

	Unit	Labour grade	Labour hours	Materials m2	Nails kg
65 x 150 x 4000mm	nr	LA	0.60	-	-
65 x 175 x 4000mm	nr	LA	0.70	-	-
65 x 200 x 4000mm	nr	LA	0.80	-	-
65 x 225 x 4000mm	nr	LA	0.90	-	-
65 x 250 x 6000mm	nr	LA	1.20	-	-
65 x 275 x 6000mm	nr	LA	1.30	-	-
65 x 300 x 6000mm	nr	LA	1.40	-	-
90 x 150 x 4000mm	nr	LA	1.00	-	-
90 x 175 x 4000mm	nr	LA	1.15	-	-
90 x 200 x 4000mm	nr	LA	1.30	-	-
90 x 225 x 4000mm	nr	LA	1.45	-	-

	Unit	Labour grade	Labour hours	Materials m2	Nails kg
90 x 250 x 6000mm	nr	LA	1.60	-	-
90 x 275 x 6000mm	nr	LA	1.75	-	-
90 x 300 x 6000mm	nr	LA	1.90	-	-
115 x 150 x 4000mm	nr	LA	2.00	-	-
115 x 175 x 4000mm	nr	LA	2.15	-	-
115 x 200 x 4000mm	nr	LA	2.30	-	-
115 x 225 x 4000mm	nr	LA	2.45	-	-
115 x 250 x 6000mm	nr	LA	2.60	-	-
115 x 275 x 6000mm	nr	LA	2.75	-	-
115 x 300 x 6000mm	nr	LA	3.00	-	-
140 x 350 x 4000mm	nr	LA	3.50	-	-
140 x 350 x 6000mm	nr	LA	3.60	-	-
140 x 350 x 8000mm	nr	LA	3.70	-	-
140 x 400 x 6000mm	nr	LA	3.80	-	-
140 x 400 x 8000mm	nr	LA	3.90	-	-
140 x 400 x 10000mm	nr	LA	4.00	-	-

Decking

Chipboard, standard grade
fixed to timber joists,
thickness

15mm	m2	LA	0.70	1.05	0.060
18mm	m2	LA	0.80	1.05	0.060
25mm	m2	LA	0.90	1.05	0.060

Plywood
fixed to timber joists,
thickness

15mm	m2	LA	1.05	1.05	0.060
18mm	m2	LA	1.10	1.05	0.060
25mm	m2	LA	1.15	1.05	0.060

	Unit	Labour grade	Labour hours	Length m	Nails kg

Joinery

Timber board flooring

Butt jointed boarding to joists, size

	Unit	Labour grade	Labour hours	Length m	Nails kg
19 x 125mm	m2	LA	0.75	8.60	0.32
19 x 100mm	m2	LA	0.80	10.75	0.38
25 x 100mm	m2	LA	0.85	8.60	0.38
25 x 150mm	m2	LA	0.80	6.60	0.30

Tongued and grooved boarding to joists, size

19 x 125mm	m2	LA	1.15	8.60	0.32
19 x 100mm	m2	LA	1.20	10.75	0.38
25 x 100mm	m2	LA	1.25	8.60	0.38
25 x 150mm	m2	LA	1.20	6.60	0.30

Chipboard boarding to floors and roofs, thickness

18mm	m2	LA	0.44	-	0.10
25mm	m2	LA	0.50	-	0.10

	Unit	Labour grade	Labour hours	Materials m2

Timber windows

Standard softwood windows without glazing bars, type

N0TV, 488 x 750mm	nr	LA	0.75	-
N09V, 488 x 900mm	nr	LA	1.00	-
N12V, 488 x 1200mm	nr	LA	1.25	-

	Unit	Labour grade	Labour hours	Materials m2
107C, 630 x 750mm	nr	LA	0.75	-
110C, 630 x 1050mm	nr	LA	1.00	-
112C, 600 x 1200mm	nr	LA	1.25	-
110T, 630 x 1050mm	nr	LA	0.75	-
113T, 630 x 1350mm	nr	LA	1.00	-
109V, 630 x 900mm	nr	LA	1.25	-
112V, 630 x 1200mm	nr	LA	0.75	-
212DG, 1200 x 1200mm	nr	LA	1.00	-
212W, 1200 x 1200mm	nr	LA	1.25	-
210C, 1200 x 1050mm	nr	LA	1.50	-
212T, 1200 x 1200mm	nr	LA	1.50	-
212CV, 1200 x 1200mm	nr	LA	1.75	-
310CVC, 1770 x 1050mm	nr	LA	1.75	-
413CWC, 2339 x 1350mm	nr	LA	1.75	-

Standard hardwood
windows without
glazing bars, type

	Unit	Labour grade	Labour hours	Materials m2
H2N10W, 915 x 1050mm	nr	LA	1.00	-
H2N13W, 915 x 1350mm	nr	LA	1.20	-
H2N15W, 915 x 1500mm	nr	LA	1.50	-
H210DG, 1200 x 1050mm	nr	LA	1.50	-
H213W, 1200 x 1350mm	nr	LA	1.60	-
H21SW, 1200 x 1500mm	nr	LA	1.70	-
H209CV, 1200 x 900mm	nr	LA	1.50	-
H213CV, 1200 x 1300mm	nr	LA	1.60	-
H309CC, 1770 x 900mm	nr	LA	2.00	-
H312CC, 1770 x 1200mm	nr	LA	2.10	-
H307C, 1770 x 750mm	nr	LA	2.00	-
H312C, 1770 x 1200mm	nr	LA	2.10	-

	Unit	Labour grade	Labour hours	Materials m2
Timber doors				
Standard flush door plywood faced both sides, 35mm thick, size				
686 x 1981mm	nr	LA	1.00	-
762 x 1981mm	nr	LA	1.00	-
Standard flush door plywood faced both sides, 40mm thick, size				
626 x 2040mm	nr	LA	1.20	-
726 x 2040mm	nr	LA	1.20	-
826 x 2040mm	nr	LA	1.20	-
Standard flush door sapele faced both sides, 35mm thick, size				
686 x 1981mm	nr	LA	1.00	-
762 x 1981mm	nr	LA	1.00	-
Standard flush door sapele faced both sides, 40mm thick, size				
626 x 2040mm	nr	LA	1.20	-
726 x 2040mm	nr	LA	1.20	-
826 x 2040mm	nr	LA	1.20	-

	Unit	Labour grade	Labour hours	Materials m2
Standard flush door teak faced both sides, 35mm thick, size				
686 x 1981mm	nr	LA	1.00	-
762 x 1981mm	nr	LA	1.00	-
Standard flush door teak faced both sides, 40mm thick, size				
626 x 2040mm	nr	LA	1.20	-
726 x 2040mm	nr	LA	1.20	-
826 x 2040mm	nr	LA	1.20	-
Standard flush door, half hour fire check, plywood faced both sides, 44mm thick, size				
686 x 1981mm	nr	LA	1.20	-
762 x 1981mm	nr	LA	1.20	-
726 x 2040mm	nr	LA	1.20	-
826 x 2040mm	nr	LA	1.20	-
Standard flush door, half hour fire check, sapele faced both sides, 44mm thick, size				
686 x 1981mm	nr	LA	1.20	-
762 x 1981mm	nr	LA	1.20	-
626 x 2040mm	nr	LA	1.20	-
726 x 2040mm	nr	LA	1.20	-
826 x 2040mm	nr	LA	1.20	-

	Unit	Labour grade	Labour hours	Materials m2
Framed ledged and braced door 44mm thick with 19mm matchboarding				
686 x 1981mm	nr	LA	1.30	-
762 x 1981mm	nr	LA	1.30	-
External hardwood panelled door 44mm thick, size				
726 x 1981mm	nr	LA	2.20	-
838 x 2040mm	nr	LA	2.20	-
Door lining size 32 x 114mm with loose stops for door, size				
686 x 1981mm	nr	LA	0.75	-
762 x 1981mm	nr	LA	0.85	-
838 x 1981mm	nr	LA	0.95	-
Door lining size 32 x 140mm with loose stops for door, size				
686 x 1981mm	nr	LA	0.75	-
762 x 1981mm	nr	LA	0.85	-
838 x 1981mm	nr	LA	0.95	-
Door frame size 38 x 114mm rebated for door, size 762 x 1981mm	nr	LA	1.10	-

	Unit	Labour grade	Labour hours	Materials m2
Door frame size 38 x 140mm rebated for door, size 762 x 1981mm	nr	LA	1.15	-

	Unit	Labour grade	Labour hours	Materials m2	Nails kg
Unframed isolated trims/skirtings					
Wrought softwood					
Architraves, skirtings					
19 x 50mm	m	LA	0.14	-	0.01
19 x 63mm	m	LA	0.14	-	0.01
25 x 50mm	m	LA	0.15	-	0.01
25 x 63mm	m	LA	0.15	-	0.01
25 x 75mm	m	LA	0.15	-	0.01
25 x 100mm	m	LA	0.17	-	0.01
25 x 125mm	m	LA	0.18	-	0.01
25 x 150mm	m	LA	0.18	-	0.01
Rails, moulded					
19 x 50mm	m	LA	0.14	-	0.01
19 x 75mm	m	LA	0.14	-	0.01
19 x 100mm	m	LA	0.15	-	0.01
25 x 50mm	m	LA	0.15	-	0.01
25 x 75mm	m	LA	0.15	-	0.01
25 x 100mm	m	LA	0.17	-	0.01
Handrail, mopstick					
50 x 50mm	m	LA	0.15	-	0.01

	Unit	Labour grade	Labour hours	Materials m2	Nails kg
Glazing beads					
13 x 25mm	m	LA	0.10	-	0.01
19 x 36mm	m	LA	0.10	-	0.01
19 x 50mm	m	LA	0.10	-	0.01
Shelving worktops **19mm thick, width**					
150mm	m	LA	0.33	-	0.01
225mm	m	LA	0.40	-	0.01
Shelving bearers					
19 x 50mm	m	LA	0.10	-	0.01
25 x 50mm	m	LA	0.10	-	0.01
Wrought hardwood					
Architraves, skirtings chamfered					
19 x 50mm	m	LA	0.21	-	0.01
19 x 63mm	m	LA	0.21	-	0.01
25 x 50mm	m	LA	0.22	-	0.01
25 x 63mm	m	LA	0.22	-	0.01
25 x 75mm	m	LA	0.22	-	0.01
25 x 100mm	m	LA	0.25	-	0.01
25 x 125mm	m	LA	0.26	-	0.01
25 x150mm	m	LA	0.26	-	0.01
Rails, moulded					
19 x 50mm	m	LA	0.21	-	0.01
19 x 75mm	m	LA	0.21	-	0.01
19 x 100mm	m	LA	0.22	-	0.01
25 x 50mm	m	LA	0.22	-	0.01
25 x 75mm	m	LA	0.22	-	0.01
25 x 100mm	m	LA	0.25	-	0.01

	Unit	Labour grade	Labour hours	Materials m2	Nails kg
Handrail, mopstick					
50 x 50mm	m	LA	0.20	-	0.01
Glazing beads					
13 x 25mm	m	LA	0.15	-	0.01
19 x 36mm	m	LA	0.15	-	0.01
19 x 50mm	m	LA	0.15	-	0.01
Shelving worktops 19mm thick, width					
150mm	m	LA	0.47	-	0.01
225mm	m	LA	0.55	-	0.01
300mm	m	LA	0.60	-	0.01
Shelving bearers					
19 x 50mm	m	LA	0.27	-	0.01
25 x 50mm	m	LA	0.25	-	0.01
Sheet linings and casings over 300mm wide					
Hardboard					
3.2mm thick	m2	LA	0.53	-	-
6.4mm thick	m2	LA	0.55	-	-
Teak faced blockboard					
18mm thick	m2	LA	0.60	-	-
Chipboard					
12mm thick	m2	LA	0.50	-	-
15mm thick	m2	LA	0.61	-	-

	Unit	Labour grade	Labour hours	Materials m2	Nails kg
Plywood					
4mm thick	m2	LA	0.34	-	-
6mm thick	m2	LA	0.36	-	-
9mm thick	m2	LA	0.40	-	-
12mm thick	m2	LA	0.46	-	-
Melamine faced chipboard 15mm	m2	LA	0.58	-	-
Insulation board					
12.7mm thick	m2	LA	0.36	-	-
19mm thick	m2	LA	0.38	-	-
25mm thick	m2	LA	0.40	-	-

Sheet linings and casings
100- 300mm wide

	Unit	Labour grade	Labour hours	Materials m2	Nails kg
Hardboard					
3.2mm thick	m	LA	0.21	-	-
6.4mm thick	m	LA	0.21	-	-
Teak faced blockboard					
18mm thick	m	LA	0.24	-	-
Chipboard					
12mm thick	m	LA	0.19	-	-
15mm thick	m	LA	0.21	-	-
Plywood					
4mm thick	m	LA	0.15	-	-
6mm thick	m	LA	0.16	-	-
9mm thick	m	LA	0.17	-	-
12mm thick	m	LA	0.19	-	-

	Unit	Labour	Labour	Materials	Nails
Melamine faced chipboard 15mm	m	LA	0.24	-	-
Insulation board					
12.7mm thick	m	LA	0.17	-	-
19mm thick	m	LA	0.18	-	-
25mm thick	m	LA	0.20	-	-

Stairs

	Unit	Labour	Labour	Materials	Nails
Wrought softwood closed tread staircase with 13 treads, 2700mm going, width 850mm, rise 2600mm	nr	LA	11.50	-	-
Wrought softwood closed tread staircase with 7 treads, 1350mm going, width 850mm, rise 1450mm	nr	LA	6.00	-	-
Landing comprising 25mm tongued and grooved flooring with rounded nosing and bearers	m2	LA	1.50	-	-
32 x 225mm wall strings	m	LA	0.50	-	-
32 x 225mm outer strings	m	LA	0.60	-	-
65 x 75mm rounded handrail	m	LA	0.45	-	-
100 x 100mm newel posts	m	LA	0.75	-	-
65 x 150 x 150mm newel caps	nr	LA	0.25	-	-

	Unit	Labour grade	Labour hours	Materials nr	Screws nr
Ironmongery					
Fixing to softwood					
Casement stay and pin	nr	LA	0.35	-	2
Casement fastener	nr	LA	0.35	-	2
Hat and coat hook	nr	LA	0.10	-	2
Shelf bracket	nr	LA	0.35	-	6
Push plate	nr	LA	0.15	-	6
Kicking plate	nr	LA	0.20	-	6
Sliding door gear					
top track	m	LA	0.30	-	4
bottom channel	m	LA	0.30	-	4
close ends	nr	LA	0.25	-	2
open bracket	nr	LA	0.25	-	2
bottom guide	nr	LA	0.25	-	2
door stop	nr	LA	0.25	-	2
top runner	nr	LA	0.33	-	2
Steel hinges					
medium butts	pr	LA	0.33	-	16
heavy butts	pr	LA	0.35	-	16
rising butts	pr	LA	0.35	-	16
friction hinges	pr	LA	0.50	-	16
Tee band hinges					
150 to 300mm	pr	LA	0.80	-	18
350 to 600mm	pr	LA	1.30	-	20

	Unit	Labour grade	Labour hours	Materials nr	Screws nr
Barrel or tower bolts					
100 to 300mm	nr	LA	0.55	-	6
350 to 450mm	nr	LA	0.60	-	6
Helical door spring	nr	LA	0.75	-	6
Overhead door spring					
medium	nr	LA	1.00	-	6
heavy	nr	LA	1.00	-	6
Door spring					
single action	nr	LA	1.75	-	6
double action	nr	LA	2.00	-	6
heavy	nr	LA	1.00	-	6
Postal knocker and letter plate	nr	LA	1.00	-	4
Pull handles	nr	LA	0.25	-	4
Flush pull handle	nr	LA	0.40	-	2
Suffolk/Norfolk latch	nr	LA	0.70	-	6
Hasp and staple	nr	LA	0.25	-	4
Flush bolts					
100 to 300mm	nr	LA	1.20	-	6
300 to 450mm	nr	LA	1.80	-	8
Indicating bolts	nr	LA	0.60	-	6

	Unit	Labour grade	Labour hours	Materials nr	Screws nr
Panic bolts					
single door	nr	LA	2.20	-	18
double door	nr	LA	3.40	-	24
Cylinder rim night latch	nr	LA	1.00	-	6
Cylinder mortice night latch	nr	LA	1.25	-	6
Rim dead lock	nr	LA	0.75	-	6
Rebated mortice lock	nr	LA	1.00	-	6
Mortice latch	nr	LA	1.20	-	4
Mortice sliding door lock	nr	LA	0.90	-	4
Mortice latch, furniture	nr	LA	0.30	-	4
Fixing to hardwood					
Casement stay and pin	nr	LA	0.45	-	2
Casement fastener	nr	LA	0.45	-	2
Hat and coat hook	nr	LA	0.15	-	2
Shelf bracket	nr	LA	0.45	-	6
Push plate	nr	LA	0.20	-	6
Kicking plate	nr	LA	0.25	-	6

	Unit	Labour grade	Labour hours	Materials nr	Screws nr
Sliding door gear					
top track	m	LA	0.40	-	4
bottom channel	m	LA	0.40	-	4
close ends	nr	LA	0.30	-	2
open bracket	nr	LA	0.30	-	2
bottom guide	nr	LA	0.30	-	2
door stop	nr	LA	0.30	-	2
top runner	nr	LA	0.35	-	2
Steel hinges					
medium butts	pr	LA	0.35	-	16
heavy butts	pr	LA	0.40	-	16
rising butts	pr	LA	0.40	-	16
friction hinges	pr	LA	0.60	-	16
Tee band hinges					
150 to 300mm	pr	LA	1.00	-	18
350 to 600mm	pr	LA	1.50	-	20
Barrel or tower bolts					
100 to 300mm	nr	LA	0.70	-	6
350 to 450mm	nr	LA	0.80	-	6
Helical door spring	nr	LA	0.90	-	6
Overhead door spring					
medium	nr	LA	1.25	-	6
heavy	nr	LA	1.25	-	6
Door spring					
single action	nr	LA	2.00	-	6
double action	nr	LA	2.25	-	6
heavy	nr	LA	1.50	-	6

	Unit	Labour grade	Labour hours	Materials nr	Screws nr
Postal knocker and letter plate	nr	LA	1.20	-	4
Pull handles	nr	LA	0.30	-	4
Flush pull handle	nr	LA	0.50	-	2
Suffolk/Norfolk latch	nr	LA	0.90	-	6
Hasp and staple	nr	LA	0.30	-	4
Flush bolts					
100 to 300mm	nr	LA	1.40	-	6
300 to 450mm	nr	LA	2.00	-	8
Indicating bolts	nr	LA	0.80	-	6
Panic bolts					
single door	nr	LA	2.50	-	18
double door	nr	LA	3.75	-	24
Cylinder rim night latch	nr	LA	1.20	-	6
Cylinder mortice night latch	nr	LA	1.50	-	6
Rim dead lock	nr	LA	0.90	-	6
Rebated mortice lock	nr	LA	1.20	-	6
Mortice latch	nr	LA	1.40	-	4
Mortice sliding door lock	nr	LA	1.00	-	4
Mortice latch, furniture	nr	LA	0.40	-	4

8

Structural steelwork

Sizes and weights of structural steel sections

Universal beams mm	kg/m	Universal beams mm	kg/m	Universal columns mm	kg/m
914 x 419	388	305 x 165	54	356 x 406	634
	343		46		551
914 x 305	289		40		467
	253	305 x 127	48		393
	224		42		340
	201		37		287
838 x 292	226	305 x 102	33		235
	194		28	356 x 368	202
	176		25		177
762 x 267	197	254 x 146	43		153
	173		37		129
	147		31	305 x 305	283
686 x 254	170	254 x 102	28		240
	152		25		198
	140		22		158
	125	203 x 133	30		137
610 x 305	238		25		118
	179	203 x 102	23		97
	149	178 x 102	19	254 x 254	167
610 x 229	140	152 x 89	16		132
	125	127 x 76	13		107
	113				89

Universal beams

mm	kg/m
533 x 210	122
	109
	101
	92
	82
457 x 191	98
	89
	82
	74
	67
457 x 52	82
	74
	67
	60
	52
406 x 178	74
	67
	60
	54
406 x 140	46
	39
356 x 171	67
	57
	51
	45

Joists

mm	kg/m
254 x 203	81.85
254 x 114	37.20
203 x 152	52.09
152 x 127	37.20
127 x 114	29.76
	26.79
127 x 76	16.37
114 x 114	26.79
102 x 102	23.07
102 x 44	7.44
89 x 89	19.35
76 x 76	14.67
	12.65

Universal columns

mm	kg/m
203 x 203	86
	71
	60
	52
	46
152 x 152	37
	30
	23

Channels

mm	kg/m
432 x 102	65.54
381 x 102	55.10
305 x 102	46.18

Tees cut from universal beams

mm	kg/m
229 x 305	70
	63
	57
	51
210 x 267	61

Tees cut from universal columns

mm	kg/m
406 x 178	118
368 x 178	101
	89
	77
	65

Channels

mm	kg/m
305 x 89	41.69
254 x 89	35.74
254 x 76	28.29
229 x 89	32.76
229 x 76	26.06
203 x 89	29.78
203 x 76	23.82
178 x 89	26.81
178 x 76	20.84
152 x 89	23.84
152 x 76	17.88
127 x 64	14.90
102 x 51	10.42
76 x 38	6.70

Tees cut from universal beams

mm	kg/m
	55
	51
	46
	41
191 x 229	49
	45
	41
	37
	34
152 x 229	41
	37
	34
	30
	26
178 x 203	37
	34
	30
	27
140 x 203	23
	20
171 x 178	34
	29
	26
	23
127 x 178	20

Tees cut from universal columns

mm	kg/m
305 x 152	79
	69
	59
	49
254 x 127	66
	54
	45
	37
203 x 102	43
	36
	30
	26
	23
152 x 76	19
	15
	12

Tees cut from universal beams

mm	kg/m
305 x 457	127
	101

Tees cut from universal beams

mm	kg/m
165 x 152	27
	23
	20

Rolled tees

mm	kg/m
51 x 51	6.92
	4.76
44 x 44	4.11
	3.14

Tees cut from universal beams mm	kg/m	Tees cut from universal beams mm	kg/m
292 x 419	113	127 x 152	24
	97		21
	88		19
267 x 381	99	102 x 152	17
	87		14
	74		13
254 x 343	85	146 x 127	22
	76		19
	70		16
	63	102 x 127	14
305 x 305	119		13
	90		11
	75	133 x 102	15
			13

Equal angles mm	kg/m	Unequal angles mm	kg/m
250 x 250 x 35	128.00	200 x 150 x 18	47.10
32	118.00	15	39.60
28	104.00	12	32.00
25	93.60	200 x 100 x 15	33.70
200 x 200 x 24	71.10	12	27.30
20	59.90	10	23.00
16	48.50	150 x 90 x 15	26.60
150 x 150 x 18	40.10	12	21.60
15	33.80	10	18.20
12	27.30	150 x 75 x 15	24.80
10	23.00	12	20.20
120 x 120 x 15	26.60	10	17.00
12	21.60	125 x 75 x 12	17.80
10	18.20	10	15.00
8	14.70	8	12.20

Equal angles mm	kg/m	Unequal angles mm	kg/m
120 x 120 x 15	26.60		
12	21.60	125 x 75 x 12	17.80
10	18.20	10	15.00
8	14.70	8	12.20
100 x 100 x 15	21.90	100 x 75 x 12	15.40
12	17.80	10	13.00
90 x 90 x 12	15.90	100 x 65 x10	12.30
10	13.48	8	9 94
8	10.90	80 x 80 x 6	8.34
7	9.61	7	7.36
6	8.30	6	6.37
80 x 80 x 10	11.90	75 x 50 x 8	7.39
8	9.63	6	5.65
6	7.34	65 x 50 x 8	6.75
70 x 70 x 10	10.30	6	5.16
8	8.36	5	4.35
6	6.38	60 x 30 x 6	3.99
60 x 60 x 10	8.69	5	3.37
8	7.09	40 x 25 x 4	1.93
6	5.42		
5	4.57		
50 x 50 x 8	5.82		
6	4.47		
5	3.77		
4	3.06		
3	2.33		
45 x 45 x 6	4.00		
5	3.38		
4	2.74		
3	2.09		
40 x 40 x 6	3.52		
5	2.97		
4	2.42		
3	1.84		

Labour grades

1 Steel erector and 1 labourer   LH

Plant grades

Mobile crane   PN

	Unit	Labour grade	Labour hours	Plant grade	Plant hours
Universal beams fixed at 3m above ground level					
203 x 133mm	t	LH	15.00	PN	3.00
254 x 146mm	t	LH	14.00	PN	2.80
305 x 127mm	t	LH	13.00	PN	2.60
305 x 165mm	t	LH	12.00	PN	2.40
406 x 178mm	t	LH	11.00	PN	2.20
457 x 191mm	t	LH	10.00	PN	2.00
Universal beams fixed at 6m above ground level					
203 x 133mm	t	LH	17.00	PN	4.00
254 x 146mm	t	LH	16.00	PN	3.80
305 x 127mm	t	LH	15.00	PN	3.60
305 x 165mm	t	LH	14.00	PN	3.40
406 x 178mm	t	LH	13.00	PN	3.20
457 x 191mm	t	LH	12.00	PN	3.00
Rolled steel joists fixed at 3m above ground level					
127 x 76mm	t	LH	24.00	PN	4.00
152 x 76mm	t	LH	23.00	PN	3.75
178 x 102mm	t	LH	22.00	PN	3.50
203 x 102mm	t	LH	21.00	PN	3.25
254 x 118mm	t	LH	20.00	PN	3.00

	Unit	Labour grade	Labour hours	Plant grade	Plant hours
Rolled steel joists fixed at 6m above ground level					
127 x 76mm	t	LH	26.00	PN	6.00
152 x 76mm	t	LH	25.00	PN	5.75
178 x 102mm	t	LH	24.00	PN	5.50
203 x 102mm	t	LH	21.00	PN	4.25
254 x 118mm	t	LH	20.00	PN	4.00
Black high strength friction grip bolts with hexagon, nut and washer					
Type M6, length					
25mm	nr	LH	0.05	-	-
50mm	nr	LH	0.06	-	-
75mm	nr	LH	0.07	-	-
100mm	nr	LH	0.08	-	-
Type M8, length					
25mm	nr	LH	0.06	-	-
50mm	nr	LH	0.07	-	-
75mm	nr	LH	0.08	-	-
100mm	nr	LH	0.09	-	-
125mm	nr	LH	0.10	-	-
Type M10, length					
50mm	nr	LH	0.09	-	-
75mm	nr	LH	0.10	-	-
100mm	nr	LH	0.11	-	-
125mm	nr	LH	0.12	-	-

	Unit	Labour grade	Labour hours	Plant grade	Plant hours
Type M12, length					
50mm	nr	LH	0.10	-	-
75mm	nr	LH	0.11	-	-
100mm	nr	LH	0.12	-	-
125mm	nr	LH	0.12	-	-
150mm	nr	LH	0.13	-	-
175mm	nr	LH	0.14	-	-
200mm	nr	LH	0.15	-	-
225mm	nr	LH	0.16	-	-
250mm	nr	LH	0.17	-	-
275mm	nr	LH	0.18	-	-
300mm	nr	LH	0.19	-	-
Type M16, length					
50mm	nr	LH	0.11	-	-
75mm	nr	LH	0.12	-	-
100mm	nr	LH	0.13	-	-
125mm	nr	LH	0.14	-	-
150mm	nr	LH	0.15	-	-
175mm	nr	LH	0.16	-	-
200mm	nr	LH	0.17	-	-
225mm	nr	LH	0.18	-	-
250mm	nr	LH	0.19	-	-
275mm	nr	LH	0.20	-	-
300mm	nr	LH	0.21	-	-

	Unit	Labour grade	Labour hours	Plant grade	Plant hours
Type M20, length					
50mm	nr	LH	0.12	-	-
75mm	nr	LH	0.13	-	-
100mm	nr	LH	0.14	-	-
125mm	nr	LH	0.15	-	-
150mm	nr	LH	0.16	-	-
175mm	nr	LH	0.17	-	-
200mm	nr	LH	0.18	-	-
225mm	nr	LH	0.19	-	-
250mm	nr	LH	0.20	-	-
275mm	nr	LH	0.21	-	-
300mm	nr	LH	0.22	-	-
Type M24, length					
50mm	nr	LH	0.13	-	-
75mm	nr	LH	0.14	-	-
100mm	nr	LH	0.15	-	-
125mm	nr	LH	0.16	-	-
150mm	nr	LH	0.17	-	-
175mm	nr	LH	0.18	-	-
200mm	nr	LH	0.19	-	-
225mm	nr	LH	0.20	-	-
250mm	nr	LH	0.21	-	-
275mm	nr	LH	0.22	-	-
300mm	nr	LH	0.23	-	-

9

Metalwork

Weights of materials

	Size mm	kg/m
Square steel bars	6	0.283
	8	0.503
	10	0.784
	12	0.130
	16	2.010
	20	3.139
	25	4.905
	32	8.035
	40	12.554
	50	19.617

	Diameter mm	kg/m
Round steel bars	6	0.222
	8	0.395
	10	0.616
	12	0.888
	16	1.579
	20	2.466
	25	3.854
	32	6.313
	40	9.864
	50	15.413

	Section mm	kg/m
Flat steel bars	25 x 9.53	1.910
	38 x 9.53	2.840
	50 x 12.70	5.060
	50 x 19.00	7.590

Labour grades

Craftsman   LA

	Unit	Labour grade	Labour hours	Materials kg
Bars				
Flat section steel bars				
25 x 10mm	m	LA	0.15	1.91
50 x 10mm	m	LA	0.18	3.80
65 x 10mm	m	LA	0.22	4.75
100 x 10mm	m	LA	0.25	7.59
Equal angle bars				
50 x 50 x 4mm	m	LA	0.25	3.06
50 x 50 x 8mm	m	LA	0.28	5.82
70 x 70 x 6mm	m	LA	0.30	6.38
70 x 70 x 10mm	m	LA	0.35	10.30
100 x 100 x 12mm	m	LA	0.45	17.80
150 x 150 x 15mm	m	LA	0.55	33.80
Lintels				
Combined galvanised steel lintels and cavity tray fixed in cavity walls, overall height 143mm, length				
750mm	nr	LA	0.15	-
900mm	nr	LA	0.17	-
1050mm	nr	LA	0.20	-
1200mm	nr	LA	0.22	-
1350mm	nr	LA	0.25	-
1500mm	nr	LA	0.28	-
1650mm	nr	LA	0.30	-
1800mm	nr	LA	0.32	-
1950mm	nr	LA	0.35	-
2100mm	nr	LA	0.38	-
2250mm	nr	LA	0.40	-
2400mm	nr	LA	0.42	-
2550mm	nr	LA	0.45	-

	Unit	Labour grade	Labour hours	Materials kg
Combined galvanised steel lintels and cavity tray fixed in cavity walls, overall height 219mm, length				
2250mm	nr	LA	0.40	-
2400mm	nr	LA	0.42	-
2550mm	nr	LA	0.45	-
2700mm	nr	LA	0.48	-
2850mm	nr	LA	0.50	-
3000mm	nr	LA	0.52	-
3300mm	nr	LA	0.55	-
3600mm	nr	LA	0.58	-
3900mm	nr	LA	0.60	-
4200mm	nr	LA	0.62	-
4575mm	nr	LA	0.65	-
4800mm	nr	LA	0.68	-
Galvanised steel lintel for internal wall 75mm wide, length				
900mm	nr	LA	0.15	-
1050mm	nr	LA	0.18	-
1200mm	nr	LA	0.20	-
Galvanised steel lintel for internal wall 100mm wide, length				
900mm	nr	LA	0.18	-
1050mm	nr	LA	0.20	-
1200mm	nr	LA	0.22	-

10

Floor finishes

Weights of materials	kg/m2
Woodblock flooring	
softwood	12.70
hardwood	17.60
Screed, 12.5mm thick	29.00
Terrazzo, 25mm thick	45.50

Mixes per m3	Cement t	Sand m3
Screed (1:3)	0.52	1.35

Tiles per m2	nr
150 x 150mm	44.36
200 x 200mm	25.00
300 x 300mm	11.09
500 x 500mm	4.00

Labour grade

Craftsman	L A

	Unit	Labour grade	Labour hours	Materials m3

Screeds

Cement and sand (1:3) beds
in floors, level and to falls
not exceeding 15 degrees
from horizontal, thickness

	Unit	Labour grade	Labour hours	Materials m3
25mm, width over 300mm	m2	LA	0.21	0.025
25mm, width not exceeding 300mm	m	LA	0.08	0.008
32mm, width over 300mm	m2	LA	0.23	0.032
32mm, width not exceeding 300mm	m	LA	0.09	0.011
38mm, width over 300mm	m2	LA	0.25	0.038
38mm, width not exceeding 300mm	m	LA	0.10	0.013
50mm, width over 300mm	m2	LA	0.29	0.050
50mm, width not exceeding 300mm	m	LA	0.12	0.017
63mm, width over 300mm	m2	LA	0.33	0.063
63mm, width not exceeding 300mm	m	LA	0.13	0.021

	Unit	Labour grade	Labour hours	Materials m3
Cement and sand (1:3) beds in floors, level and to falls exceeding 15 degrees from horizontal, thickness				
25mm, width over 300mm	m2	LA	0.30	0.025
25mm, width not exceeding 300mm	m	LA	0.12	0.008
32mm, width over 300mm	m2	LA	0.33	0.032
32mm, width not exceeding 300mm	m	LA	0.13	0.011
38mm, width over 300mm	m2	LA	0.35	0.038
38mm, width not exceeding 300mm	m	LA	0.14	0.013
50mm, width over 300mm	m2	LA	0.40	0.050
50mm, width not exceeding 300mm	m	LA	0.15	0.017
63mm, width over 300mm	m2	LA	0.55	0.063
63mm, width not exceeding 300mm	m	LA	0.16	0.021

	Unit	Labour grade	Labour hours	Materials m3
Granolithic, cement and granite chippings (1:2:5) in beds in floors, level and to falls not exceeding 15 degrees from horizontal, steel trowelled smooth, thickness				
25mm, width over 300mm	m2	LA	0.38	0.025
25mm, width not exceeding 300mm	m	LA	0.15	0.008
32mm, width over 300mm	m2	LA	0.42	0.032
32mm, width not exceeding 300mm	m	LA	0.16	0.011
38mm, width over 300mm	m2	LA	0.45	0.038
38mm, width not exceeding 300mm	m	LA	0.18	0.013
50mm, width over 300mm	m2	LA	0.48	0.050
50mm, width not exceeding 300mm	m	LA	0.19	0.017
63mm, width over 300mm	m2	LA	0.52	0.063
63mm, width not exceeding 300mm	m	LA	0.21	0.021

	Unit	Labour grade	Labour hours	Materials m3
Granolithic, cement and granite chippings (1:2:5) in beds in floors, level and to falls exceeding 15 degrees from horizontal, steel trowelled smooth, thickness				
25mm, width over 300mm	m2	LA	0.50	0.025
25mm, width not exceeding 300mm	m	LA	0.18	0.008
32mm, width over 300mm	m2	LA	0.55	0.032
32mm, width not exceeding 300mm	m	LA	0.22	0.011
38mm, width over 300mm	m2	LA	0.60	0.038
38mm, width not exceeding 300mm	m	LA	0.25	0.013
50mm, width over 300mm	m2	LA	0.65	0.050
50mm, width not exceeding 300mm	m	LA	0.28	0.017
63mm, width over 300mm	m2	LA	0.70	0.063
63mm, width not exceeding 300mm	m	LA	0.30	0.021

	Unit	Labour grade	Labour hours	Materials m3
Lightweight concrete beds, cement and exfoliated vermiculite (1:8), not exceeding 15 degrees from horizontal, screeded finish, thickness				
25mm, width over 300mm	m2	LA	0.21	0.025
25mm, width not exceeding 300mm	m	LA	0.08	0.008
38mm, width over 300mm	m2	LA	0.25	0.038
38mm, width not exceeding 300mm	m	LA	0.10	0.013
50mm, width over 300mm	m2	LA	0.29	0.050
50mm, width not exceeding 300mm	m	LA	0.12	0.017
75mm, width over 300mm	m2	LA	0.40	0.075
75mm, width not exceeding 300mm	m	LA	0.16	0.025

	Unit	Labour grade	Labour hours	Materials m3
Lightweight concrete beds, cement and exfoliated vermiculite (1:8), exceeding 15 degrees from horizontal, screeded finish, thickness				
25mm, width over 300mm	m2	LA	0.30	0.025
25mm, width not exceeding 300mm	m	LA	0.12	0.008
38mm, width over 300mm	m2	LA	0.35	0.038
38mm, width not exceeding 300mm	m	LA	0.16	0.013
50mm, width over 300mm	m2	LA	0.40	0.050
50mm, width not exceeding 300mm	m	LA	0.18	0.017
75mm, width over 300mm	m2	LA	0.50	0.075
75mm, width not exceeding 300mm	m	LA	0.20	0.025

	Unit	Labour grade	Labour hours	Tiles	Mortar m3

Tiling

Red clay quarry floor tiles, bedded on 12mm cement mortar (1:3), to falls not exceeding 15 degrees, butt joints straight both ways

	Unit	Labour grade	Labour hours	Tiles	Mortar m3
150 x 150 x 12.5mm thick over 300mm wide	m2	LA	0.90	44.36	0.013
150 x 150 x 12.5mm thick not exceeding 300mm wide	m	LA	0.35	8.33	0.004
200 x 200 x 12.5mm thick over 300mm wide	m2	LA	0.80	44.36	0.013
200 x 200 x 12.5mm thick not exceeding 300mm wide	m	LA	0.32	8.33	0.004

Red clay quarry floor tiles, bedded on 12mm cement mortar (1:3), to falls exceeding 15 degrees, butt joints straight both ways

	Unit	Labour grade	Labour hours	Tiles	Mortar m3
150 x 150 x 12.5mm thick over 300mm wide	m2	LA	1.00	44.36	0.013
150 x 150 x 12.5mm thick not exceeding 300mm wide	m	LA	0.40	8.33	0.004
200 x 200 x 12.5mm thick over 300mm wide	m2	LA	0.80	44.36	0.013
200 x 200 x 12.5mm thick not exceeding 300mm wide	m	LA	0.35	8.33	0.004

	Unit	Labour grade	Labour hours	Tiles	Mortar m3
Vitrified plain ceramic floor tiles, bedded on 12mm cement mortar (1:3), to falls not exceeding 15 degrees, pointing with matching grout					
150 x 150 x 12.5mm thick over 300mm wide	m2	LA	0.80	44.36	0.013
150 x 150 x 12.5mm thick not exceeding 300mm wide	m	LA	0.35	8.33	0.004
Vitrified plain ceramic floor tiles, bedded on 12mm cement mortar (1:3), to falls exceeding 15 degrees, pointing with matching grout					
150 x 150 x 12.5mm thick over 300mm wide	m2	LA	0.90	44.36	0.013
150 x 150 x 12.5mm thick not exceeding 300mm wide	m	LA	0.40	8.33	0.004
Terrazzo floor tiles, bedded on 12mm cement mortar (1:3), to falls not exceeding 15 degrees, pointing in white cement					
300 x 300 x 25mm thick over 300mm wide	m2	LA	1.50	11.11	0.013
300 x 300 x 25mm thick not exceeding 300mm wide	m	LA	0.55	3.70	0.004

	Unit	Labour grade	Labour hours	Tiles nr
Rubber floor tiles, size 500 x 500mm, to falls not exceeding 15 degrees, fixed with adhesive				
over 300mm wide	m2	LA	0.40	4.00
not exceeding 300mm wide	m	LA	0.14	1.33
Rubber floor tiles, size 500 x 500mm, to falls not exceeding 15 degrees, fixed with adhesive				
over 300mm wide	m2	LA	0.50	4.00
not exceeding 300mm wide	m	LA	0.16	1.33
Thermoplastic floor tiles, size 300 x 300mm, to falls not exceeding 15 degrees, fixed with adhesive				
over 300mm wide	m2	LA	0.25	3.70
not exceeding 300mm wide	m	LA	0.10	1.23
Thermoplastic floor tiles, size 300 x 300mm, to falls exceeding 15 degrees, fixed with adhesive				
over 300mm wide	m2	LA	0.30	3.70
not exceeding 300mm wide	m	LA	0.12	1.23

	Unit	Labour grade	Labour hours	Tiles nr
Vinyl floor tiles, size 300 x 300mm, to falls not exceeding 15 degrees, fixed with adhesive				
over 300mm wide	m2	LA	0.25	3.70
not exceeding 300mm wide	m	LA	0.10	1.23
Vinyl floor tiles, size 300 x 300mm, to falls exceeding 15 degrees, fixed with adhesive				
over 300mm wide	m2	LA	0.30	3.70
not exceeding 300mm wide	m	LA	0.12	1.23
Cork floor tiles, size 300 x 300mm, to falls not exceeding 15 degrees, fixed with adhesive				
over 300mm wide	m2	LA	0.30	3.70
not exceeding 300mm wide	m	LA	0.12	1.23
Cork floor tiles, size 300 x 300mm, to falls exceeding 15 degrees, fixed with adhesive				
over 300mm wide	m2	LA	0.35	3.70
not exceeding 300mm wide	m	LA	0.14	1.23

	Unit	Labour grade	Labour hours	Tiles nr
Polypropylene carpet tiles with bitumen backing, size 300 x 300mm, to falls not exceeding 15 degrees, laid loose				
over 300mm wide	m2	LA	0.20	3.70
not exceeding 300mm wide	m	LA	0.08	1.23
Polypropylene carpet tiles with bitumen backing, size 300 x 300mm, to falls exceeding 15 degrees, laid loose				
over 300mm wide	m2	LA	0.28	3.70
not exceeding 300mm wide	m	LA	0.10	1.23
Hardwood block flooring, size 225 x 75 x 25mm thick, to falls not exceeding 15 degrees, laid herringbone pattern				
over 300mm wide	m2	LA	1.80	3.70
not exceeding 300mm wide	m	LA	0.75	1.23
Hardwood block flooring, size 225 x 75 x 25mm thick, to falls exceeding 15 degrees, laid herringbone pattern				
over 300mm wide	m2	LA	1.80	3.70
not exceeding 300mm wide	m	LA	0.85	1.23

	Unit	Labour grade	Labour hours	Tiles nr
Sheeting				
Vinyl sheet flooring to falls not exceeding 15 degrees, fixed with adhesive				
over 300mm wide	m2	LA	0.20	-
not exceeding 300mm wide	m	LA	0.07	-
Vinyl sheet flooring to falls exceeding 15 degrees, fixed with adhesive				
over 300mm wide	m2	LA	0.22	-
not exceeding 300mm wide	m	LA	0.08	-
Carpet				
Polypropylene edge-fitted latex-backed fitted carpet to falls not exceeding 15 degrees				
over 300mm wide	m2	LA	0.20	-
not exceeding 300mm wide	m	LA	0.06	-
Polypropylene edge-fitted latex-backed fitted carpet to falls exceeding 15 degrees				
over 300mm wide	m2	LA	0.28	-
not exceeding 300mm wide	m	LA	0.08	-

	Unit	Labour grade	Labour hours	Tiles nr

In situ flooring

In situ terrazzo flooring, white cement and white aggregate, to falls not exceeding 15 degrees, thickness

	Unit	Labour grade	Labour hours	Tiles nr
16mm, width over 300mm	m2	LA	2.00	-
16mm, width not exceeding 300mm	m2	LA	0.80	-

In situ terrazzo flooring, white cement and white aggregate, to falls exceeding 15 degrees, thickness

	Unit	Labour grade	Labour hours	Tiles nr
16mm, width over 300mm wide	m2	LA	2.10	-
16mm, width not exceeding 300mm	m2	LA	0.90	-

11

Wall and ceiling finishes, partitions

Weights of materials	kg/m3
Cement	1440
Lime, hydrated	500
Sand	1600
	kg/m2
Carlite browning, 11mm thick	7.80
Carlite tough coat, 11mm thick	7.80
Carlite bonding coat	
8mm thick	7.10
11mm thick	9.80
Thistle hardwall, 11mm thick	8.80
Thistle dri-coat, 11mm thick	8.30
Thistle renovating, 11mm thick	8.80
Thistle universal one coat, 13mm thick	12.00

Coverage	Thickness mm	m2/1000kg
Carlite browning	11	135-155
Carlite tough coat	11	135-150
Carlite bonding	11	100-115
Carlite hardwall	11	115-130
Thistle dri-coat	11	125-135
Thistle renovating	11	115-125
Thistle universal	13	85-95

Tile coverings (per m2)

Size mm	nr
152 x 152	43.27
200 x 200	25.00

Wallboard sizes

Thickness mm	Width mm	Length mm
9.5	600	1800
9.5	900	1800
9.5	1200	1800
12.5	600	2286
12.5	600	2350
12.5	600	2400
12.5	600	2438
12.5	600	2700
12.5	600	3000
12.5	900	2286
12.5	900	2350
12.5	900	2400
12.5	900	2438

Thickness mm	Width mm	Length mm
12.5	900	2700
12.5	900	3000
12.5	1200	2286
12.5	1200	2350
12.5	1200	2400
12.5	1200	2438
12.5	1200	2700
12.5	1200	3000
12.5	1200	3300
12.5	1200	3600

Labour grades

2 Plasterers and 1 labourer   LA

	Unit	Labour grade	Labour hours	Nails kg
Baseboard, wallboard square edge, 1200 x 2400mm x 9.5mm thick, taped butt joints, to receive skim coat, to timber with nails				
Walls, height 2.10 to 2.40m	m	LA	0.55	2.50
Walls, height 2.40 to 2.70m	m	LA	0.65	2.80
Walls, height 2.70 to 3.00m	m	LA	0.75	3.10
Walls, height 3.00 to 3.30m	m	LA	0.85	3.40
Walls, height 3.30 to 3.60m	m	LA	0.95	3.70
Ceilings	m2	LA	0.30	1.00
Sides and soffits of beams, girth not exceeding 600mm	m	LA	0.20	0.04
Sides and soffits of beams, girth 600 to 1200mm	m	LA	0.35	0.08
Sides and soffits of beams, girth 1200 to 1800mm	m	LA	0.55	0.12
Sides of columns, girth not exceeding 600mm	m	LA	0.18	0.04
Sides of columns, girth 600 to 1200mm	m	LA	0.33	0.08
Sides of columns, girth 1200 to 1800mm	m	LA	0.50	0.12
Reveals, openings and recesses not exceeding 300mm wide	m	LA	0.20	0.02
Reveals, openings and recesses 300 to 600mm wide	m	LA	0.30	0.04

	Unit	Labour grade	Labour hours	Nails kg
Baseboard, wallboard square edge, 1200 x 2400mm x 12.5mm thick, taped butt joints, to receive skim coat, to timber with nails				
Walls, height 2.10 to 2.40m	m	LA	0.65	2.50
Walls, height 2.40 to 2.70m	m	LA	0.75	2.80
Walls, height 2.70 to 3.00m	m	LA	0.85	3.10
Walls, height 3.00 to 3.30m	m	LA	0.95	3.40
Walls, height 3.30 to 3.60m	m	LA	1.05	3.70
Ceilings	m2	LA	0.38	1.00
Sides and soffits of beams, girth not exceeding 600mm	m	LA	0.28	0.04
Sides and soffits of beams, girth 600 to 1200mm	m	LA	0.42	0.08
Sides and soffits of beams, girth 1200 to 1800mm	m	LA	0.64	0.12
Sides of columns, girth not exceeding 600mm	m	LA	0.20	0.04
Sides of columns, girth 600 to 1200mm	m	LA	0.38	0.08
Sides of columns, girth 1200 to 1800mm	m	LA	0.55	0.12
Reveals, openings and recesses not exceeding 300mm wide	m	LA	0.25	0.02
Reveals, openings and recesses 300 to 600mm wide	m	LA	0.35	0.04

	Unit	Labour grade	Labour hours	Nails kg
Baseboard, square edge plank, size 600 x 2400 x 19mm, taped butt joints, to receive skim coat, to timber with nails				
Walls, height 2.10 to 2.40m	m	LA	0.75	2.50
Walls, height 2.40 to 2.70m	m	LA	0.85	2.80
Walls, height 2.70 to 3.00m	m	LA	0.95	3.10
Walls, height 3.00 to 3.30m	m	LA	1.05	3.40
Walls, height 3.30 to 3.60m	m	LA	1.15	3.70
Ceilings	m2	LA	0.38	1.00
Sides and soffits of beams, girth not exceeding 600mm	m	LA	0.33	0.04
Sides and soffits of beams, girth 600 to 1200mm	m	LA	0.45	0.08
Sides and soffits of beams, girth 1200 to 1800mm	m	LA	0.68	0.12
Sides of columns, girth not exceeding 600mm	m	LA	0.25	0.04
Sides of columns, girth 600 to 1200mm	m	LA	0.42	0.08
Sides of columns, girth 1200 to 1800mm	m	LA	0.60	0.12
Reveals, openings and recesses not exceeding 300mm wide	m	LA	0.25	0.02
Reveals, openings and recesses 300 to 600mm wide	m	LA	0.35	0.04

	Unit	Labour grade	Labour hours	Nails kg
Thermal board 25mm thick with tapered edges, joint filler and taped joints, to receive skim coat, to timber with nails				
Walls, height 2.10 to 2.40m	m	LA	0.85	2.80
Walls, height 2.40 to 2.70m	m	LA	0.95	2.80
Walls, height 2.70 to 3.00m	m	LA	1.05	3.10
Walls, height 3.00 to 3.30m	m	LA	1.15	3.40
Walls, height 3.30 to 3.60m	m	LA	1.25	3.70
Ceilings	m2	LA	0.40	1.00
Sides and soffits of beams, girth not exceeding 600mm	m	LA	0.35	0.04
Sides and soffits of beams, girth 600 to 1200mm	m	LA	0.48	0.08
Sides and soffits of beams, girth 1200 to 1800mm	m	LA	0.70	0.12
Sides of columns, girth not exceeding 600mm	m	LA	0.28	0.04
Sides of columns, girth 600 to 1200mm	m	LA	0.45	0.08
Sides of columns, girth 1200 to 1800mm	m	LA	0.65	0.12
Reveals, openings and recesses not exceeding 300mm wide	m	LA	0.30	0.02
Reveals, openings and recesses 300 to 600mm wide	m	LA	0.40	0.04

	Unit	Labour grade	Labour hours	Nails kg
Thermal board 32mm thick with tapered edges, joint filler and taped joints, to receive skim coat, to timber with nails				
Walls, height 2.10 to 2.40m	m	LA	0.90	2.80
Walls, height 2.40 to 2.70m	m	LA	0.95	2.80
Walls, height 2.70 to 3.00m	m	LA	1.00	3.10
Walls, height 3.00 to 3.30m	m	LA	1.10	3.40
Walls, height 3.30 to 3.60m	m	LA	1.15	3.70
Ceilings	m2	LA	0.45	1.00
Sides and soffits of beams, girth not exceeding 600mm	m	LA	0.38	0.04
Sides and soffits of beams, girth 600 to 1200mm	m	LA	0.50	0.08
Sides and soffits of beams, girth 1200 to 1800mm	m	LA	0.72	0.12
Sides of columns, girth not exceeding 600mm	m	LA	0.30	0.04
Sides of columns, girth 600 to 1200mm	m	LA	0.48	0.08
Sides of columns, girth 1200 to 1800mm	m	LA	0.68	0.12
Reveals, openings and recesses not exceeding 300mm wide	m	LA	0.32	0.02
Reveals, openings and recesses 300 to 600mm wide	m	LA	0.42	0.04

	Unit	Labour grade	Labour hours	Nails kg
Thermal board 40mm thick with tapered edges, joint filler and taped joints, to receive skim coat, to timber with nails				
Walls, height 2.10 to 2.40m	m	LA	0.95	2.80
Walls, height 2.40 to 2.70m	m	LA	1.00	2.80
Walls, height 2.70 to 3.00m	m	LA	1.05	3.10
Walls, height 3.00 to 3.30m	m	LA	1.05	3.40
Walls, height 3.30 to 3.60m	m	LA	1.10	3.70
Ceilings	m2	LA	0.50	1.00
Sides and soffits of beams, girth not exceeding 600mm	m	LA	0.40	0.04
Sides and soffits of beams, girth 600 to 1200mm	m	LA	0.52	0.08
Sides and soffits of beams, girth 1200 to 1800mm	m	LA	0.74	0.12
Sides of columns, girth not exceeding 600mm	m	LA	0.32	0.04
Sides of columns, girth 600 to 1200mm	m	LA	0.50	0.08
Sides of columns, girth 1200 to 1800mm	m	LA	0.70	0.12
Reveals, openings and recesses not exceeding 300mm wide	m	LA	0.34	0.02
Reveals, openings and recesses 300 to 600mm wide	m	LA	0.44	0.04

	Unit	Labour grade	Labour hours	Nails kg
Thermal board 50mm thick with tapered edges, joint filler and taped joints, to receive skim coat, to timber with nails				
Walls, height 2.10 to 2.40m	m	LA	1.00	2.80
Walls, height 2.40 to 2.70m	m	LA	1.05	2.80
Walls, height 2.70 to 3.00m	m	LA	1.10	3.10
Walls, height 3.00 to 3.30m	m	LA	1.15	3.40
Walls, height 3.30 to 3.60m	m	LA	1.20	3.70
Ceilings	m2	LA	0.55	1.00
Sides and soffits of beams, girth not exceeding 600mm	m	LA	0.42	0.04
Sides and soffits of beams, girth 600 to 1200mm	m	LA	0.55	0.08
Sides and soffits of beams, girth 1200 to 1800mm	m	LA	0.76	0.12
Sides of columns, girth not exceeding 600mm	m	LA	0.35	0.04
Sides of columns, girth 600 to 1200mm	m	LA	0.52	0.08
Sides of columns, girth 1200 to 1800mm	m	LA	0.72	0.12
Reveals, openings and recesses not exceeding 300mm wide	m	LA	0.36	0.02
Reveals, openings and recesses 300 to 600mm wide	m	LA	0.46	0.04

	Unit	Labour grade	Labour hours	Nails kg

Preformed dry partition, 9.5mm wallboard both sides of cellular core

57mm thick, 38 x 38mm
jointing battens, grey
faced with square butt
joints for plastering

	Unit	Labour grade	Labour hours	Nails kg
Walls, height 2.10 to 2.40m	m	LA	1.20	-
Walls, height 2.40 to 2.70m	m	LA	1.30	-
Walls, height 2.70 to 3.00m	m	LA	1.40	-
Walls, height 3.00 to 3.30m	m	LA	1.50	-
Walls, height 3.30 to 3.60m	m	LA	1.60	-

63mm thick, 38 x 38mm
jointing battens, grey
faced with square butt
joints for plastering

Walls, height 2.10 to 2.40m	m	LA	1.25	-
Walls, height 2.40 to 2.70m	m	LA	1.35	-
Walls, height 2.70 to 3.00m	m	LA	1.45	-
Walls, height 3.00 to 3.30m	m	LA	1.55	-
Walls, height 3.30 to 3.60m	m	LA	1.65	-

	Unit	Labour grade	Labour hours	Nails kg
Partitions, metal stud partition, board on metal studs				
75mm thick, 48mm wide studs at 600mm maximum centres, 12.5mm thick tapered edge wallboard both sides with joints taped				
Walls, height 2.10 to 2.40m	m	LA	1.35	-
Walls, height 2.40 to 2.70m	m	LA	1.45	-
Walls, height 2.70 to 3.00m	m	LA	1.55	-
Walls, height 3.00 to 3.30m	m	LA	1.65	-
Walls, height 3.30 to 3.60m	m	LA	1.75	-
173mm thick, 146mm wide studs at 600mm maximum centres, 12.5mm thick tapered edge wallboard both sides with joints taped				
Walls, height 2.10 to 2.40m	m	LA	1.45	-
Walls, height 2.40 to 2.70m	m	LA	1.55	-
Walls, height 2.70 to 3.00m	m	LA	1.65	-
Walls, height 3.00 to 3.30m	m	LA	1.75	-
Walls, height 3.30 to 3.60m	m	LA	1.85	-

	Unit	Labour grade	Labour hours	Browning tonne	Finish tonne

Plasterwork

Premixed lightweight plaster, 11mm browning, 2mm finish

Walls

	Unit	Labour grade	Labour hours	Browning tonne	Finish tonne
over 300mm wide	m2	LA	0.45	0.007	0.003
not exceeding 300mm wide	m2	LA	0.18	0.002	0.001

Curved walls

	Unit	Labour grade	Labour hours	Browning tonne	Finish tonne
over 300mm wide	m2	LA	0.50	0.007	0.003
not exceeding 300mm wide	m	LA	0.20	0.002	0.001

Ceilings

	Unit	Labour grade	Labour hours	Browning tonne	Finish tonne
over 300mm wide	m2	LA	0.60	0.007	0.003
not exceeding 300mm wide	m	LA	0.30	0.002	0.001
Sides and soffits of beams, girth not exceeding 600mm	m	LA	0.30	0.004	0.002
Sides and soffits of beams, girth 600 to 1200mm	m	LA	0.55	0.008	0.004
Sides and soffits of beams, girth 1200 to 1800mm	m	LA	0.80	0.013	0.005
Sides of columns, girth not exceeding 600mm	m	LA	0.25	0.004	0.002

	Unit	Labour grade	Labour hours	Browning tonne	Finish tonne
Sides of columns, girth 600 to 1200mm	m	LA	0.50	0.008	0.004
Sides of columns, girth 1200 to 1800mm	m	LA	0.70	0.013	0.005
Reveals, openings and recesses not exceeding 300mm wide	m	LA	0.34	0.002	0.001
Reveals, openings and recesses 300 to 600mm wide	m	LA	0.44	0.004	0.002

	Unit	Labour grade	Labour hours	Bonding tonne	Finish tonne
Premixed lightweight plaster, 8mm bonding, 2mm finish					
Walls					
over 300mm wide	m2	LA	0.41	0.007	0.003
not exceeding 300mm wide	m2	LA	0.16	0.002	0.001
Curved walls					
over 300mm wide	m2	LA	0.45	0.007	0.003
not exceeding 300mm wide	m	LA	0.18	0.002	0.001
Ceilings					
over 300mm wide	m2	LA	0.55	0.007	0.003
not exceeding 300mm wide	m	LA	0.28	0.002	0.001

	Unit	Labour grade	Labour hours	Bonding tonne	Finish tonne
Sides and soffits of beams, girth not exceeding 600mm	m	LA	0.28	0.004	0.002
Sides and soffits of beams, girth 600 to 1200mm	m	LA	0.45	0.008	0.004
Sides and soffits of beams, girth 1200 to 1800mm	m	LA	0.70	0.012	0.006
Sides of columns, girth not exceeding 600mm	m	LA	0.22	0.004	0.002
Sides of columns, girth 600 to 1200mm	m	LA	0.42	0.008	0.004
Sides of columns, girth 1200 to 1800mm	m	LA	0.60	0.012	0.006
Reveals, openings and recesses not exceeding 300mm wide	m	LA	0.30	0.002	0.001
Reveals, openings and recesses 300 to 600mm wide	m	LA	0.40	0.004	0.002

	Unit	Labour grade	Labour hours	Finish tonne
One coat finishing plaster, 2mm thick				
Walls				
over 300mm wide	m2	LA	0.20	0.0025
not exceeding 300mm wide	m2	LA	0.08	0.0008

	Unit	Labour grade	Labour hours	Finish tonne
Curved walls				
over 300mm wide	m2	LA	0.35	0.0025
not exceeding 300mm wide	m	LA	0.12	0.0008
Ceilings				
over 300mm wide	m2	LA	0.30	0.0025
not exceeding 300mm wide	m	LA	0.10	0.0008
Sides and soffits of beams, girth not exceeding 600mm	m	LA	0.12	0.0015
Sides and soffits of beams, girth 600 to 1200mm	m	LA	0.24	0.0030
Sides and soffits of beams, girth 1200 to 1800mm	m	LA	0.36	0.0045
Sides of columns, girth not exceeding 600mm	m	LA	0.10	0.0015
Sides of columns, girth 600 to 1200mm	m	LA	0.20	0.0030
Sides of columns, girth 1200 to 1800mm	m	LA	0.30	0.0045
Reveals, openings and recesses not exceeding 300mm wide	m	LA	0.10	0.0008

	Unit	Labour grade	Labour hours	Finish tonne
Reveals, openings and recesses 300-600mm wide	m	LA	0.40	0.0016

Two coats finishing plaster, 5mm thick

Walls

over 300mm wide	m	LA	0.30	0.0060
not exceeding 300mm wide	m	LA	0.10	0.0020

Curved walls

over 300mm wide	m2	LA	0.40	0.0060
not exceeding 300mm wide	m	LA	0.14	0.0020

Ceilings

over 300mm wide	m2	LA	0.35	0.0060
not exceeding 300mm wide	m	LA	0.12	0.0020
Sides and soffits of beams, girth not exceeding 600mm	m	LA	0.18	0.0036
Sides and soffits of beams, girth 600 to 1200mm	m	LA	0.36	0.0072
Sides and soffits of beams, girth 1200 to 1800mm	m	LA	0.54	0.0108

	Unit	Labour grade	Labour hours	Finish tonne
Sides of columns, girth not exceeding 600mm	m	LA	0.16	0.0036
Sides of columns, girth 600 to 1200mm	m	LA	0.32	0.0072
Sides of columns, girth 1200 to 1800mm	m	LA	0.48	0.0108
Reveals, openings and recesses not exceeding 300mm wide	m	LA	0.10	0.0020
Reveals, openings and recesses 300 to 600mm wide	m	LA	0.40	0.0040

	Unit	Labour grade	Labour hours	Mortar m3
Cement and sand (1:3) mortar 12mm thick				
Walls				
over 300mm wide	m2	LA	0.40	0.012
not exceeding 300mm wide	m	LA	0.12	0.004
Curved walls				
over 300mm wide	m2	LA	0.55	0.012
not exceeding 300mm wide	m	LA	0.15	0.004

	Unit	Labour grade	Labour hours	Mortar m3
Ceilings				
over 300mm wide	m2	LA	0.60	0.012
not exceeding 300mm wide	m	LA	0.20	0.004
Sides and soffits of beams, girth not exceeding 600mm	m	LA	0.24	0.007
Sides and soffits of beams, girth 600 to 1200mm	m	LA	0.48	0.014
Sides and soffits of beams, girth 1200 to 1800mm	m	LA	0.72	0.021
Sides of columns, girth not exceeding 600mm	m	LA	0.20	0.007
Sides of columns, girth 600 to 1200mm	m	LA	0.40	0.014
Sides of columns, girth 1200 to 1800mm	m	LA	0.60	0.021
Reveals, openings and recesses not exceeding 300mm wide	m	LA	0.20	0.004
Reveals, openings and recesses 300 to 600mm wide	m	LA	0.40	0.004

	Unit	Labour grade	Labour hours	Mortar m3
Cement and sand (1:3) mortar 19mm thick				
Walls				
over 300mm wide	m2	LA	0.70	0.019
not exceeding 300mm wide	m	LA	0.22	0.006
Curved walls				
over 300mm wide	m2	LA	0.75	0.019
not exceeding 300mm wide	m	LA	0.25	0.006
Ceilings				
over 300mm wide	m2	LA	0.80	0.019
not exceeding 300mm wide	m	LA	0.28	0.006
Sides and soffits of beams, girth not exceeding 600mm	m	LA	0.42	0.012
Sides and soffits of beams, girth 600 to 1200mm	m	LA	0.82	0.024
Sides and soffits of beams, girth 1200 to 1800mm	m	LA	1.22	0.036
Sides of columns, girth not exceeding 600mm	m	LA	0.38	0.012

	Unit	Labour grade	Labour hours	Mortar m3
Sides of columns, girth 600 to 1200mm	m	LA	0.75	0.024
Sides of columns, girth 1200 to 1800mm	m	LA	1.15	0.036
Reveals, openings and recesses not exceeding 300mm wide	m	LA	0.25	0.006
Reveals, openings and recesses 300 to 600mm wide	m	LA	0.50	0.012

	Unit	Labour grade	Labour hours	Tiles nr	Grout kg
Wall tiling					
Glazed ceramic wall tiles, size 150 x 150 x 6mm, fixing with adhesive, pointing with matching grout					
over 300mm wide	m2	LA	0.70	43.27	0.28
not exceeding 300mm wide	m	LA	0.22	14.32	0.09
Raking cutting	m	LA	0.07	-	-
Curved cutting	m	LA	0.07	-	-
Cut and fit around small pipes	nr	LA	0.05	-	-

	Unit	Labour grade	Labour hours	Tiles nr	Grout kg
Glazed ceramic wall tiles, size 200 x 200 x 6mm, fixing with adhesive, pointing with matching grout					
over 300mm wide	m2	LA	0.60	21.00	0.28
not exceeding 300mm wide	m	LA	0.20	7.00	0.09
Raking cutting	m	LA	0.07	-	-
Curved cutting	m	LA	0.07	-	-
Cut and fit around small pipes	nr	LA	0.05	-	-

	Unit	Labour grade	Labour hours
Partitions			
Cellular core plasterboard dry partition 57mm thick including softwood battens and taped vertical joints	m2	LA	0.60
Metal stud partition 170mm thick including metal frame, softwood sole plate and two layers of plasterboard	m2	LA	1.30
Softwood stud partition consisting sole and head plates, studs, noggings and two coats of plasterboard	m2	LA	0.70

12

Plumbing

Weights of materials

Diameter mm	Copper tubes Table X mm	Table Y mm	Table Z mm
6	0.091	0.117	0.077
8	0.125	0.162	0.105
10	0.158	0.206	0.133
12	0.191	0.251	0.161
15	0.280	0.392	0.203
18	0.385	0.476	0.292
22	0.531	0.697	0.359
28	0.681	0.899	0.459
35	1.133	1.409	0.670
42	1.368	1.700	0.922
54	1.769	2.905	1.334

Mild steel cisterns

Length mm	Width mm	Depth mm	Capacity litres
457	305	305	18
610	305	371	36
610	406	371	54
610	432	432	54
610	432	432	68
610	457	482	86
686	508	508	114
736	559	559	159
762	584	610	191
914	610	584	227
914	660	610	264

Plastic cisterns

Ref.	Capacity litres	Capacity gallons	Weight kg
PC4	18	4	0.85
PC15	68	15	2.95
PC25	114	25	3.40
PC40	182	40	6.35

Roof drainage	**Area m2**	**Pipe mm**	**Gutter mm**
One end outlet	15	50	75
	38	68	100
	100	110	150
Centre outlet	30	50	75
	75	68	100
	200	110	150

Jointing materials (per joint)	**Lead kg**	**Yarn kg**
Cast iron soil pipes		
50mm	0.65	0.07
75mm	1.10	0.10
100mm	1.85	0.13

Labour grade

Craftsman   LA

	Unit	Labour grade	Labour hours	Brackets nr
Rainwater pipes				
68mm diameter PVC-U pipe fixed to brick walling	m	LA	0.25	0.50
shoe	nr	LA	0.30	-
bend	nr	LA	0.15	-
branch	nr	LA	0.20	-
100mm diameter PVC-U pipe fixed to brick walling	m	LA	0.28	0.50
shoe	nr	LA	0.32	-
bend	nr	LA	0.16	-
branch	nr	LA	0.22	-
65 x 65mm square PVC-U pipe fixed to brick walling	m	LA	0.25	0.50
shoe	nr	LA	0.30	-
bend	nr	LA	0.15	-
branch	nr	LA	0.20	-
63mm diameter aluminium pipe fixed to brick walling	m	LA	0.28	0.50
shoe	nr	LA	0.33	-
bend	nr	LA	0.18	-
branch	nr	LA	0.23	-
76mm diameter aluminium pipe fixed to brick walling	m	LA	0.33	0.50
shoe	nr	LA	0.38	-
bend	nr	LA	0.23	-
branch	nr	LA	0.28	-

	Unit	Labour grade	Labour hours	Brackets nr
102mm diameter aluminium pipe fixed to brick walling	m	LA	0.36	0.50
shoe	nr	LA	0.38	-
bend	nr	LA	0.28	-
branch	nr	LA	0.31	-
50mm diameter cast iron pipe fixed to brick walling	m	LA	0.25	0.50
shoe	nr	LA	0.30	-
bend	nr	LA	0.15	-
branch	nr	LA	0.20	-
65mm diameter cast iron pipe fixed to brick walling	m	LA	0.28	0.50
shoe	nr	LA	0.33	-
bend	nr	LA	0.18	-
branch	nr	LA	0.23	-
75mm diameter cast iron pipe fixed to brick walling	m	LA	0.35	0.50
shoe	nr	LA	0.40	-
bend	nr	LA	0.25	-
branch	nr	LA	0.30	-
100mm diameter cast iron pipe fixed to brick walling	m	LA	0.40	0.50
shoe	nr	LA	0.45	-
bend	nr	LA	0.30	-
branch	nr	LA	0.35	-

	Unit	Labour grade	Labour hours	Brackets nr
Rainwater gutters				
75mm PVC-U gutter fixed to timber	m	LA	0.23	0.50
outlet	nr	LA	0.23	-
stop end	nr	LA	0.12	-
angle	nr	LA	0.23	-
110mm PVC-U gutter fixed to timber	m	LA	0.26	0.50
outlet	nr	LA	0.26	-
stop end	nr	LA	0.14	-
angle	nr	LA	0.26	-
100mm aluminium gutter fixed to timber	m	LA	0.29	0.50
outlet	nr	LA	0.29	-
stop end	nr	LA	0.17	-
angle	nr	LA	0.29	-
125mm aluminium gutter fixed to timber	m	LA	0.31	0.50
outlet	nr	LA	0.31	-
stop end	nr	LA	0.20	-
angle	nr	LA	0.31	-
100mm cast iron gutter fixed to timber	m	LA	0.35	0.50
outlet	nr	LA	0.35	-
stop end	nr	LA	0.20	-
angle	nr	LA	0.35	-

	Unit	Labour grade	Labour hours	Brackets nr
150mm cast iron gutter fixed to timber	m	LA	0.45	0.50
outlet	nr	LA	0.45	-
stop end	nr	LA	0.28	-
angle	nr	LA	0.45	-

Waste pipes

	Unit	Labour grade	Labour hours	Brackets nr
32mm diameter PVC-U waste pipe fixed to plastered walls	m	LA	0.24	0.50
bend	nr	LA	0.22	-
tee	nr	LA	0.25	-
40mm diameter PVC-U waste pipe fixed to plastered walls	m	LA	0.27	0.50
bend	nr	LA	0.24	-
tee	nr	LA	0.28	-
50mm diameter PVC-U waste pipe fixed to plastered walls	m	LA	0.30	0.50
bend	nr	LA	0.26	-
tee	nr	LA	0.30	-

	Unit	Labour grade	Labour hours	Brackets nr
Soil pipes				
82mm diameter PVC-U soil pipe fixed to brick walling	m	LA	0.33	0.50
bend	nr	LA	0.33	-
branch	nr	LA	0.35	-
110mm diameter PVC-U soil pipe fixed to brick walling	m	LA	0.35	0.50
bend	nr	LA	0.35	-
branch	nr	LA	0.39	-
160mm diameter PVC-U soil pipe fixed to brick walling	m	LA	0.38	0.50
bend	nr	LA	0.38	-
branch	nr	LA	0.43	-
75mm diameter cast iron soil pipe fixed to brick walling	m	LA	0.45	0.50
bend	nr	LA	0.45	-
branch	nr	LA	0.50	-
100mm diameter cast iron soil pipe fixed to brick walling	m	LA	0.55	0.50
bend	nr	LA	0.55	-
branch	nr	LA	0.60	-

	Unit	Labour grade	Labour hours	Clips nr
Overflows				
19mm PVC-U overflow pipe fixed to brick walling	m	LA	0.19	0.50
bend	nr	LA	0.19	-
branch	nr	LA	0.22	-
Traps				
Polypropylene trap screwed to outlet and pipe				
32mm	nr	LA	0.27	-
40mm	nr	LA	0.32	-

	Unit	Labour grade	Labour hours	Clips nr
Copper pipework				
Copper pipe, Table W, capillary joints fixed with clips to timber				
8mm	m	LA	0.21	0.80
10mm	m	LA	0.21	0.80
15mm	m	LA	0.21	0.80
22mm	m	LA	0.22	0.80
28mm	m	LA	0.24	0.80
35mm	m	LA	0.29	0.80
42mm	m	LA	0.32	0.80
54mm	m	LA	0.34	0.80

	Unit	Labour grade	Labour hours	Clips nr
Copper pipe, Table W, capillary joints, plugged and screwed				
8mm	m	LA	0.23	0.80
10mm	m	LA	0.23	0.80
15mm	m	LA	0.23	0.80
22mm	m	LA	0.24	0.80
28mm	m	LA	0.26	0.80
35mm	m	LA	0.31	0.80
42mm	m	LA	0.34	0.80
54mm	m	LA	0.36	0.80
Copper pipe, Table Z, capillary joints fixed with clips to timber				
15mm	m	LA	0.21	0.80
22mm	m	LA	0.22	0.80
28mm	m	LA	0.24	0.80
35mm	m	LA	0.29	0.80
42mm	m	LA	0.32	0.80
54mm	m	LA	0.34	0.80
Copper pipe, Table Z, capillary joints, plugged and screwed				
15mm	m	LA	0.23	0.80
22mm	m	LA	0.24	0.80
28mm	m	LA	0.26	0.80
35mm	m	LA	0.31	0.80
42mm	m	LA	0.34	0.80
54mm	m	LA	0.36	0.80

	Unit	Labour grade	Labour hours	Clips nr
Copper pipe, Table X, DZR compression joints fixed with clips to timber				
15mm	m	LA	0.20	0.80
22mm	m	LA	0.21	0.80
28mm	m	LA	0.23	0.80
35mm	m	LA	0.28	0.80
42mm	m	LA	0.31	0.80
54mm	m	LA	0.33	0.80
Copper pipe, Table X, DZR compression joints, plugged and screwed				
15mm	m	LA	0.22	0.80
22mm	m	LA	0.23	0.80
28mm	m	LA	0.25	0.80
35mm	m	LA	0.30	0.80
42mm	m	LA	0.33	0.80
54mm	m	LA	0.35	0.80
Copper pipe, Table Z, DZR compression joints fixed with clips to timber				
15mm	m	LA	0.20	0.80
22mm	m	LA	0.21	0.80
28mm	m	LA	0.23	0.80
35mm	m	LA	0.28	0.80
42mm	m	LA	0.31	0.80
54mm	m	LA	0.33	0.80
Copper pipe, Table Z, DZR compression joints, plugged and screwed				
15mm	m	LA	0.22	0.80
22mm	m	LA	0.23	0.80
28mm	m	LA	0.25	0.80
35mm	m	LA	0.30	0.80
42mm	m	LA	0.33	0.80

	Unit	Labour grade	Labour hours	Clips nr
Copper pipe, Table X, brass compression joints fixed with clips to timber				
15mm	m	LA	0.20	0.80
22mm	m	LA	0.21	0.80
28mm	m	LA	0.23	0.80
Copper pipe, Table X, brass compression joints, plugged and screwed				
15mm	m	LA	0.22	0.80
22mm	m	LA	0.23	0.80
28mm	m	LA	0.25	0.80
Copper pipe, Table Z, brass compression joints fixed with clips to timber				
15mm	m	LA	0.20	0.80
22mm	m	LA	0.21	0.80
28mm	m	LA	0.23	0.80
Copper pipe, Table Z, brass compression joints, plugged and screwed				
15mm	m	LA	0.22	0.80
22mm	m	LA	0.23	0.80
28mm	m	LA	0.25	0.80
Copper pipe, Table Y, DZR compression joints, in trenches				
15mm	m	LA	0.11	-
18mm	m	LA	0.12	-
22mm	m	LA	0.13	-
28mm	m	LA	0.15	-
35mm	m	LA	0.17	-
42mm	m	LA	0.18	-
54mm	m	LA	0.20	-

	Unit	Labour grade	Labour hours	Clips nr
Medium density polyethylene pipe, in trenches				
20mm	m	LA	0.11	-
25mm	m	LA	0.11	-
32mm	m	LA	0.17	-
50mm	m	LA	0.20	-
63mm	m	LA	0.21	-

	Unit	Labour grade	Labour hours
Extra over for lead-free pre-soldered capillary joints and fittings for 8mm diameter copper pipe			
Made bend	nr	LA	0.15
Straight coupling	nr	LA	0.20
Straight connector	nr	LA	0.20
Reducer	nr	LA	0.20
Elbow	nr	LA	0.20
Tee	nr	LA	0.25
Stop end	nr	LA	0.15
Extra over for lead-free pre-soldered capillary joints and fittings for 10mm diameter copper pipe			
Made bend	nr	LA	0.17
Straight coupling	nr	LA	0.22
Straight connector	nr	LA	0.22
Reducer	nr	LA	0.22
Elbow	nr	LA	0.22
Tee	nr	LA	0.27
Stop end	nr	LA	0.17

	Unit	Labour grade	Labour hours
Extra over for lead-free pre-soldered capillary joints and fittings for 15mm diameter copper pipe			
Made bend	nr	LA	0.18
Straight coupling	nr	LA	0.23
Straight connector	nr	LA	0.23
Reducer	nr	LA	0.23
Elbow	nr	LA	0.23
Tee	nr	LA	0.28
Stop end	nr	LA	0.18
Extra over for lead-free pre-soldered capillary joints and fittings for 22mm diameter copper pipe			
Made bend	nr	LA	0.20
Straight coupling	nr	LA	0.25
Straight connector	nr	LA	0.25
Reducer	nr	LA	0.25
Elbow	nr	LA	0.25
Tee	nr	LA	0.30
Stop end	nr	LA	0.20
Extra over for lead-free pre-soldered capillary joints and fittings for 28mm diameter copper pipe			
Made bend	nr	LA	0.22
Straight coupling	nr	LA	0.27
Straight connector	nr	LA	0.27
Reducer	nr	LA	0.27
Elbow	nr	LA	0.27
Tee	nr	LA	0.32
Stop end	nr	LA	0.22

	Unit	Labour grade	Labour hours

Extra over for lead-free pre-soldered capillary joints and fittings for 35mm diameter copper pipe

	Unit	Labour grade	Labour hours
Made bend	nr	LA	0.24
Straight coupling	nr	LA	0.29
Straight connector	nr	LA	0.29
Reducer	nr	LA	0.29
Elbow	nr	LA	0.29
Tee	nr	LA	0.34
Stop end	nr	LA	0.24

Extra over for lead-free pre-soldered capillary joints and fittings for 42mm diameter copper pipe

	Unit	Labour grade	Labour hours
Made bend	nr	LA	0.26
Straight coupling	nr	LA	0.31
Straight connector	nr	LA	0.31
Reducer	nr	LA	0.31
Elbow	nr	LA	0.31
Tee	nr	LA	0.36
Stop end	nr	LA	0.26

Extra over for lead-free pre-soldered capillary joints and fittings for 42mm diameter copper pipe

	Unit	Labour grade	Labour hours
Made bend	nr	LA	0.28
Straight coupling	nr	LA	0.33
Straight connector	nr	LA	0.33
Reducer	nr	LA	0.33
Elbow	nr	LA	0.33
Tee	nr	LA	0.38
Stop end	nr	LA	0.28

	Unit	Labour grade	Labour hours
Extra over for lead-free pre-soldered capillary joints and fittings for 54mm diameter copper pipe			
Made bend	nr	LA	0.30
Straight coupling	nr	LA	0.35
Straight connector	nr	LA	0.35
Reducer	nr	LA	0.35
Elbow	nr	LA	0.35
Tee	nr	LA	0.40
Stop end	nr	LA	0.30
Extra over for dezinc-ification-resistant compression fittings for 15mm diameter copper pipe			
Made bend	nr	LA	0.16
Straight coupling	nr	LA	0.21
Straight connector	nr	LA	0.21
Reducer	nr	LA	0.21
Elbow	nr	LA	0.21
Tee	nr	LA	0.26
Stop end	nr	LA	0.16
Extra over for dezinc-ification-resistant compression fittings for 22mm diameter copper pipe			
Made bend	nr	LA	0.18
Straight coupling	nr	LA	0.23
Straight connector	nr	LA	0.23
Reducer	nr	LA	0.23
Elbow	nr	LA	0.23
Tee	nr	LA	0.28
Stop end	nr	LA	0.18

	Unit	Labour grade	Labour hours
Extra over for dezinc-ification-resistant compression fittings for 28mm diameter copper pipe			
Made bend	nr	LA	0.20
Straight coupling	nr	LA	0.25
Straight connector	nr	LA	0.25
Reducer	nr	LA	0.25
Elbow	nr	LA	0.25
Tee	nr	LA	0.30
Stop end	nr	LA	0.20
Extra over for dezinc-ification-resistant compression fittings for 35mm diameter copper pipe			
Made bend	nr	LA	0.22
Straight coupling	nr	LA	0.27
Straight connector	nr	LA	0.27
Reducer	nr	LA	0.27
Elbow	nr	LA	0.27
Tee	nr	LA	0.32
Stop end	nr	LA	0.22
Extra over for dezinc-ification-resistant compression fittings for 42mm diameter copper pipe			
Made bend	nr	LA	0.24
Straight coupling	nr	LA	0.29
Straight connector	nr	LA	0.29
Reducer	nr	LA	0.29
Elbow	nr	LA	0.29
Tee	nr	LA	0.34
Stop end	nr	LA	0.24

	Unit	Labour grade	Labour hours
Extra over for dezinc-ification-resistant compression fittings for 54mm diameter copper pipe			
Made bend	nr	LA	0.26
Straight coupling	nr	LA	0.31
Straight connector	nr	LA	0.31
Reducer	nr	LA	0.31
Elbow	nr	LA	0.31
Tee	nr	LA	0.36
Stop end	nr	LA	0.26
Stopcocks and valves			
Gunmetal stopcock with brass headwork, capillary joints copper to copper			
15mm	nr	LA	0.20
22mm	nr	LA	0.25
28mm	nr	LA	0.30
Gunmetal stopcock with brass headwork, compression joints copper to copper			
15mm	nr	LA	0.18
22mm	nr	LA	0.23
28mm	nr	LA	0.28
Radiator valve with compression fitting to taper male union outlet, 15mm	nr	LA	0.35

	Unit	Labour grade	Labour hours

Insulation

Rigid mineral glass fibre
sectional pipe lagging
19mm thick, secured with
aluminium bands to
pipework, diameter

15mm	nr	LA	0.06
22mm	nr	LA	0.08
28mm	nr	LA	0.10
35mm	nr	LA	0.12
42mm	nr	LA	0.14
54mm	nr	LA	0.16

Polyethylene glass fibre
insulation jacket 60mm
thick to cold water cisterns,
size

450 x 300 x 300mm	nr	LA	0.50
600 x 500 x 400mm	nr	LA	0.70
675 x 525 x 525mm	nr	LA	0.90
1000 x 625 x 525mm	nr	LA	1.10

Polyethylene glass fibre
insulation jacket 80mm
thick to hot water cylinder,
size

400mm diameter x 900mm high	nr	LA	0.50
450mm diameter x 900mm high	nr	LA	0.60
500mm diameter x 1050mm high	nr	LA	0.70

	Unit	Labour grade	Labour hours

Tanks and cylinders

Galvanised steel cold water
tanks with cover, capacity

	Unit	Labour grade	Labour hours
18 litres	nr	LA	0.70
36 litres	nr	LA	0.80
54 litres	nr	LA	0.90
68 litres	nr	LA	1.00
86 litres	nr	LA	1.10
114 litres	nr	LA	1.20
154 litres	nr	LA	1.30
191 litres	nr	LA	1.40
227 litres	nr	LA	1.50
260 litres	nr	LA	1.60
330 litres	nr	LA	1.70
414 litres	nr	LA	1.80

Direct copper hot water
cylinders, capacity

	Unit	Labour grade	Labour hours
98 litres	nr	LA	0.70
120 litres	nr	LA	1.20
148 litres	nr	LA	1.30
166 litres	nr	LA	1.50

Indirect copper hot water
cylinders, capacity

	Unit	Labour grade	Labour hours
96 litres	nr	LA	0.90
114 litres	nr	LA	1.40
140 litres	nr	LA	1.50
162 litres	nr	LA	1.70

	Unit	Labour grade	Labour hours
Sanitary fittings			
Bath, 1700mm long, with waste, overflow, chain and plug			
Acrylic	nr	LA	3.00
Vitreous enamelled	nr	LA	3.25
Cast iron	nr	LA	3.50
Bath panel			
End	nr	LA	0.25
Panel	nr	LA	0.35
Vitreous china wash basin with waste, overflow, chain and plug, size			
560 x 430mm	nr	LA	2.00
590 x 440mm	nr	LA	2.10
Stainless steel sink unit, single bowl and single drainer with waste, overflow, chain and plug, overall size 1000 x 500mm	nr	LA	1.70
Stainless steel sink unit, single bowl and double drainer with waste, overflow, chain and plug, overall size 1500 x 500mm	nr	LA	1.90
Stainless steel sink unit, double bowl and double drainer with waste, overflow, chain and plug, overall size 1500 x 500mm	nr	LA	2.00
Vitreous china free-standing plain rimmed bidet	nr	LA	2.00

	Unit	Labour grade	Labour hours
Low level WC suite with cistern, pan, ball valve, flush pipe, overflow, seat and cover	nr	LA	3.00

Taps

Chromium plated pillar taps

13mm	nr	LA	0.35
19mm	nr	LA	0.40

Chromium plated pillar deck pattern sink mixer tap	nr	LA	0.75

13

Glazing

Weights of materials	Thickness m m	kg/m2	Maximum pane size mm
Float glass	3	7.50	6000 x 3180
	4	10.00	6000 x 3180
	5	12.50	6000 x 3180
	6	15.00	6000 x 3180
	10	25.00	6000 x 3180
	12	30.00	6000 x 3180
	15	37.50	4600 x 3180
	19	47.50	4600 x 3180
	25	63.50	4600 x 3180
Clear sheet glass	3	7.50	2130 x 1230
	4	10.00	2760 x 1220
	5	12.50	2130 x 2400
	6	15.00	2130 x 2400
Patterned glass	3	6.00	2140 x 1320
	4	7.50	2140 x 1320
	5	9.50	2140 x 1320
	6	11.50	3200 x 1320
	10	21.50	3200 x 1320

Putty per m2	Wood kg/m2	Metal kg/m2
Panes up 0.10m2	3.95	5.08
	4.21	5.42
	4.56	5.86
	4.96	6.38
	5.33	6.86
Panes 0.10 - 0.50m2	1.62	2.09
	1.72	2.21
	1.87	2.41
	2.03	2.61
	2.18	2.80
Panes 0.50 - 1.00m2	1.03	1.32
	1.09	1.40
	1.19	1.52
	1.28	1.64
	1.38	1.77
Panes over 1.00m2	0.73	0.93
	0.74	0.99
	0.77	1.08
	0.91	1.17
	0.97	1.25

Labour grades

Craftsman   LA

	Unit	Labour grade	Labour hours	Putty kg	Beads m
Clear float glass					
In wood with putty and sprigs under 0.15m2, thickness					
3mm	m2	LA	0.90	1.00	-
4mm	m2	LA	0.90	1.00	-
5mm	m2	LA	0.90	1.00	-
6mm	m2	LA	1.00	1.00	-
10mm	m2	LA	1.05	1.00	-
In wood with putty and sprigs over 0.15m2, thickness					
3mm	m2	LA	0.60	0.26	-
4mm	m2	LA	0.60	0.26	-
5mm	m2	LA	0.60	0.26	-
6mm	m2	LA	0.65	0.26	-
10mm	m2	LA	0.70	0.26	-
In wood with pinned beads under 0.15m2, thickness					
3mm	m2	LA	1.20	-	0.95
4mm	m2	LA	1.20	-	0.95
5mm	m2	LA	1.20	-	0.95
6mm	m2	LA	1.35	-	0.95
10mm	m2	LA	1.50	-	0.95

	Unit	Labour grade	Labour hours	Putty kg	Beads m
In wood with pinned beads over 0.15m2, thickness					
3mm	m2	LA	0.80	-	0.40
4mm	m2	LA	0.80	-	0.40
5mm	m2	LA	0.80	-	0.40
6mm	m2	LA	0.90	-	0.40
10mm	m2	LA	1.00	-	0.40
In metal with putty under 0.15m2, thickness					
3mm	m2	LA	1.20	1.60	-
4mm	m2	LA	1.20	1.60	-
5mm	m2	LA	1.20	1.60	-
6mm	m2	LA	1.35	1.60	-
10mm	m2	LA	1.50	1.60	-
In metal with putty over 0.15m2, thickness					
3mm	m2	LA	0.80	0.40	-
4mm	m2	LA	0.80	0.40	-
5mm	m2	LA	0.80	0.40	-
6mm	m2	LA	0.90	0.40	-
10mm	m2	LA	1.00	0.40	-

	Unit	Labour grade	Labour hours	Putty kg	Beads m
In metal with clipped beads under 0.15m2, thickness					
3mm	m2	LA	1.50	-	0.95
4mm	m2	LA	1.50	-	0.95
5mm	m2	LA	1.50	-	0.95
6mm	m2	LA	1.65	-	0.95
10mm	m2	LA	1.80	-	0.95
In metal with clipped beads over 0.15m2, thickness					
3mm	m2	LA	1.00	-	0.40
4mm	m2	LA	1.00	-	0.40
5mm	m2	LA	1.00	-	0.40
6mm	m2	LA	1.10	-	0.40
10mm	m2	LA	1.20	-	0.40

White patterned glass

	Unit	Labour grade	Labour hours	Putty kg	Beads m
In wood with putty and sprigs under 0.15m2, thickness					
4mm	m2	LA	0.90	1.00	-
5mm	m2	LA	0.90	1.00	-
6mm	m2	LA	1.00	1.00	-
In wood with putty and sprigs over 0.15m2, thickness					
4mm	m2	LA	0.60	0.26	-
5mm	m2	LA	0.60	0.26	-
6mm	m2	LA	0.65	0.26	-

	Unit	Labour grade	Labour hours	Putty kg	Beads m
In wood with pinned beads under 0.15m2, thickness					
4mm	m2	LA	1.20	-	0.95
5mm	m2	LA	1.20	-	0.95
6mm	m2	LA	1.35	-	0.95
In wood with pinned beads over 0.15m2, thickness					
4mm	m2	LA	0.80	-	0.40
5mm	m2	LA	0.80	-	0.40
6mm	m2	LA	0.90	-	0.40
In metal with putty under 0.15m2, thickness					
4mm	m2	LA	1.20	1.60	-
5mm	m2	LA	1.20	1.60	-
6mm	m2	LA	1.35	1.60	-
In metal with putty over 0.15m2, thickness					
4mm	m2	LA	0.80	0.40	-
5mm	m2	LA	0.80	0.40	-
6mm	m2	LA	0.90	0.40	-

	Unit	Labour grade	Labour hours	Putty kg	Beads m
In metal with clipped beads under 0.15m2, thickness					
4mm	m2	LA	1.50	-	0.95
5mm	m2	LA	1.50	-	0.95
6mm	m2	LA	1.65	-	0.95
In metal with clipped beads over 0.15m2, thickness					
4mm	m2	LA	1.00	-	0.40
5mm	m2	LA	1.00	-	0.40
6mm	m2	LA	1.10	-	0.40
In metal with putty under 0.15m2, thickness					
4mm	m2	LA	1.20	1.60	-
5mm	m2	LA	1.20	1.60	-
6mm	m2	LA	1.35	1.60	-

Rough cast glass

	Unit	Labour grade	Labour hours	Putty kg	Beads m
In wood with putty and sprigs under 0.15m2, thickness					
6mm	m2	LA	1.00	1.60	-
10mm	m2	LA	1.05	1.60	-

	Unit	Labour grade	Labour hours	Putty kg	Beads m
In wood with putty and sprigs over 0.15m2, thickness					
6mm	m2	LA	0.65	0.40	-
10mm	m2	LA	0.70	0.40	-
In wood with pinned beads under 0.15m2, thickness					
6mm	m2	LA	1.35	-	0.95
10mm	m2	LA	1.50	-	0.95
In wood with pinned beads over 0.15m2, thickness					
6mm	m2	LA	0.90	-	0.40
10mm	m2	LA	1.10	-	0.40
In wood with screwed beads under 0.15m2, thickness					
6mm	m2	LA	1.80	-	0.95
10mm	m2	LA	1.95	-	0.95
In wood with screwed beads over 0.15m2, thickness					
6mm	m2	LA	1.20	-	0.40
10mm	m2	LA	1.30	-	0.40

	Unit	Labour grade	Labour hours	Putty kg	Beads m
In metal with putty under 0.15m2, thickness					
6mm	m2	LA	1.35	1.60	-
10mm	m2	LA	1.50	1.60	-
In metal with putty over 0.15m2, thickness					
6mm	m2	LA	0.80	0.40	-
10mm	m2	LA	1.00	0.40	-
In metal with clipped beads under 0.15m2, thickness					
6mm	m2	LA	1.65	-	0.95
10mm	m2	LA	1.80	-	0.95
In metal with clipped beads over 0.15m2, thickness					
6mm	m2	LA	1.10	-	0.40
10mm	m2	LA	1.20	-	0.40

Georgian wired cast glass

	Unit	Labour grade	Labour hours	Putty kg	Beads m
In wood with putty and sprigs under 0.15m2, thickness					
7mm	m2	LA	1.00	1.00	-

	Unit	Labour grade	Labour hours	Putty kg	Beads m
In wood with putty and sprigs over 0.15m2, thickness 7mm	m2	LA	0.65	0.26	-
In wood with pinned beads and sprigs under 0.15m2, thickness 7mm	m2	LA	1.35	-	0.95
In wood with pinned beads and sprigs over 0.15m2, thickness 7mm	m2	LA	0.90	-	0.40
In wood with screwed beads and sprigs under 0.15m2, thickness 7mm	m2	LA	1.80	-	0.95
In wood with screwed beads and sprigs over 0.15m2, thickness 7mm	m2	LA	1.20	-	0.40
In metal with putty under 0.15m2, thickness 7mm	m2	LA	1.35	1.60	-
In metal with putty over 0.15m2, thickness 7mm	m2	LA	0.90	0.40	-
In metal with clipped beads under 0.15m2, thickness 7mm	m2	LA	1.65	-	0.95

	Unit	Labour grade	Labour hours	Putty kg	Beads m
In metal with clipped beads over 0.15m2, thickness 7mm	m2	LA	1.65	-	0.40
In metal with putty over 0.15m2, thickness 7mm	m2	LA	0.90	0.40	-
In metal with clipped beads over 0.15m2, thickness 7mm	m2	LA	0.90	0.40	-

Anti-sun float glass

In wood with putty and sprigs under 0.15m2, thickness

	Unit	Labour grade	Labour hours	Putty kg	Beads m
4mm	m2	LA	0.90	1.00	-
6mm	m2	LA	1.00	1.00	-
10mm	m2	LA	1.05	1.00	-
12mm	m2	LA	1.20	1.00	-

In wood with putty and sprigs over 0.15m2, thickness

	Unit	Labour grade	Labour hours	Putty kg	Beads m
4mm	m2	LA	0.60	0.40	-
6mm	m2	LA	0.65	0.40	-
10mm	m2	LA	0.70	0.40	-
12mm	m2	LA	0.80	0.40	-

	Unit	Labour grade	Labour hours	Putty kg	Beads m
In wood with pinned beads under 0.15m2, thickness					
4mm	m2	LA	1.20	-	0.95
6mm	m2	LA	1.35	-	0.95
10mm	m2	LA	1.50	-	0.95
12mm	m2	LA	1.80	-	0.95
In wood with pinned beads over 0.15m2, thickness					
4mm	m2	LA	0.80	-	0.40
6mm	m2	LA	0.90	-	0.40
10mm	m2	LA	1.00	-	0.40
12mm	m2	LA	1.20	-	0.40
In wood with screwed beads under 0.15m2, thickness					
4mm	m2	LA	1.65	-	0.95
6mm	m2	LA	1.80	-	0.95
10mm	m2	LA	1.95	-	0.95
12mm	m2	LA	2.25	-	0.95
In wood with screwed beads over 0.15m2, thickness					
4mm	m2	LA	1.10	-	0.40
6mm	m2	LA	1.20	-	0.40
10mm	m2	LA	1.30	-	0.40
12mm	m2	LA	1.50	-	0.40

	Unit	Labour grade	Labour hours	Putty kg	Beads m
In metal with putty under 0.15m2, thickness					
4mm	m2	LA	1.20	1.00	-
6mm	m2	LA	1.35	1.00	-
10mm	m2	LA	1.50	1.00	-
12mm	m2	LA	1.80	1.00	-
In metal with putty over 0.15m2, thickness					
4mm	m2	LA	0.80	0.40	-
6mm	m2	LA	0.90	0.40	-
10mm	m2	LA	1.00	0.40	-
12mm	m2	LA	1.20	0.40	-
In metal with clipped beads under 0.15m2, thickness					
4mm	m2	LA	1.50	-	0.95
6mm	m2	LA	1.65	-	0.95
10mm	m2	LA	1.80	-	0.95
12mm	m2	LA	2.10	-	0.95
In metal with clipped beads over 0.15m2, thickness					
4mm	m2	LA	1.00	-	0.40
6mm	m2	LA	1.10	-	0.40
10mm	m2	LA	1.20	-	0.40
12mm	m2	LA	1.40	-	0.40

	Unit	Labour grade	Labour hours	Beads m
Toughened float glass				
In metal with gaskets and screwed metal beads in panes, maximum size				
2000 x 1200mm, 4mm thick	m2	LA	0.60	0.40
2000 x 1200mm, 5mm thick	m2	LA	0.70	0.40
2600 x 1500mm, 6mm thick	m2	LA	0.80	0.40
3960 x 1520mm, 10mm thick	m2	LA	1.10	0.40
3100 x 2500mm, 12mm thick	m2	LA	1.50	0.40
Double glazing units				
Factory manufactured hermetically sealed double glazing units consisting of two 5mm thick panes of clear float or white patterned glass fixed with beads and gaskets in prepared frames, pane size				
520 x 420mm	nr	LA	1.20	1.88
520 x 620mm	nr	LA	1.50	2.28
750 x 750mm	nr	LA	1.80	3.00
850 x 850mm	nr	LA	2.10	3.40
1050 x 1050mm	nr	LA	2.40	4.20

14

Wallpapering

Roll sizes	Length m	Width m	Area m2
UK	10.00	0.53	5.30
France	11.00	0.57	6.27
USA	7.31	0.46	3.36
Europe	10.65	0.71	7.56

Rolls required

No allowances for normal waste or for door and window areas.

Room perimeter m	Wall height m	Rolls required nr
8	2.5	4
9	2.5	5
10	2.5	5
11	2.5	6
12	2.5	6
13	2.5	7
14	2.5	7
15	2.5	8
16	2.5	8
17	2.5	8
18	2.5	9
19	2.5	10
20	2.5	10
21	2.5	10
22	2.5	11
23	2.5	11

Room perimeter m	Wall height m	Rolls required nr
24	2.5	12
26	2.5	13
27	2.5	13
28	2.5	14
8	2.8	5
9	2.8	5
10	2.8	5
11	2.8	7
12	2.8	7
13	2.8	7
14	2.8	8
15	2.8	8
16	2.8	9
17	2.8	10
18	2.8	10
19	2.8	11
20	2.8	11
21	2.8	12
22	2.8	13
23	2.8	13
24	2.8	14
26	2.8	14
27	2.8	15
28	2.8	15

Labour grade

Craftsman   LA

	Unit	Labour grade	Labour hours
Prepare, size, apply adhesive and hang lining paper to			
Walls and columns areas less than 0.5m2	nr	LA	0.30
Walls and columns areas more than 0.5m2	m2	LA	0.25
Ceilings and beams areas less than 0.5m2	nr	LA	0.35
Ceilings and beams areas more than 0.5m2	m2	LA	0.30
Prepare, size, apply adhesive and hang vinyl paper to			
Walls and columns areas less than 0.5m2	nr	LA	0.32
Walls and columns areas more than 0.5m2	m2	LA	0.27
Ceilings and beams areas less than 0.5m2	nr	LA	0.37
Ceilings and beams areas more than 0.5m2	m2	LA	0.32

	Unit	Labour grade	Labour hours
Prepare, size, apply adhesive and hang embossed paper to			
Walls and columns areas less than 0.5m2	nr	LA	0.35
Walls and columns areas more than 0.5m2	m2	LA	0.30
Ceilings and beams areas less than 0.5m2	nr	LA	0.40
Ceilings and beams areas more than 0.5m2	m2	LA	0.35
Prepare, size, apply adhesive and hang textured paper to			
Walls and columns areas less than 0.5m2	nr	LA	0.35
Walls and columns areas more than 0.5m2	m2	LA	0.30
Ceilings and beams areas less than 0.5m2	nr	LA	0.40
Ceilings and beams areas more than 0.5m2	m2	LA	0.35

	Unit	Labour grade	Labour hours
Prepare, size, apply adhesive and hang hessian paper to			
Walls and columns areas less than 0.5m2	nr	LA	0.50
Walls and columns areas more than 0.5m2	m2	LA	0.45
Ceilings and beams areas less than 0.5m2	nr	LA	0.55
Ceilings and beams areas more than 0.5m2	m2	LA	0.50
Prepare, size, apply adhesive and hang suede paper to			
Walls and columns areas less than 0.5m2	nr	LA	0.50
Walls and columns areas more than 0.5m2	m2	LA	0.45
Ceilings and beams areas less than 0.5m2	nr	LA	0.55
Ceilings and beams areas more than 0.5m2	m2	LA	0.50

	Unit	Labour grade	Labour hours
Prepare, size, apply adhesive and hang border strips to walls	m	LA	0.06

15

Painting

Average coverage of paints

The following schedule of average coverage figures for painting work is the 1974 revision of the schedule compiled and approved for the guidance of commercial organisations when assessing the values of materials in painting work by the Painting Industries' Liaison Committee (constituent bodies: British Decorators' Association, National Federation of Painting and Decorating Contractors, Paintmakers Association of Great Britain and Scottish Decorators' Federation) whose permission to publish is hereby acknowledged.

In this revision a range of spreading capacities is given. Figures are in square metres per litre, except for oil-bound water paint and cement-based paint which are stated in square metres per kilogram. Figures are given for a single coat, but users are recommended to follow manufacturers' recommendations on when to use single or multicoat systems.

It is emphasised that the figures quoted in the schedule are for brush applications achieved in scale painting work and take into account losses and wastage. They are not optimum figures based upon ideal conditions, nor minimum figures reflecting the reverse of these conditions.

There will be instances when the figures indicated by paint manufacturers in their literature will be higher than those shown in the schedule. The Committee realise that under ideal conditions of application, and depending on such factors as the skill of the applicator and the type and quality of the product, better covering figures can be achieved.

The figures stated are for application by brush and for appropriate systems on each surface. They are given for guidance and allow for variation depending on certain factors.

Average coverage of paints in square metres per litre

Surfaces	A	B	C	D	E	F	G	H
Water-thinned primer/undercoat								
primer	13-15	-	-	-	-	-	10-12	7-10
undercoat	-	-	-	-	-	-	-	10-12
Plaster primer (including building board)	9-11	8-12	9-11	7-9	5-7	2-4	8-10	7-9
Alkali-resistant primer	7-11	6-8	7-11	6-8	4-6	2-4	-	-
External wall primer sealer	6-8	6-7	6-8	5-7	4-6	2-4	-	-
Undercoat	11-14	7-9	7-9	6-8	6-8	3-4	11-14	10-12
Gloss finish	11-14	8-10	8-10	7-9	6-8	-	11-14	10-12
Oil-based thixotropic finish	Figures should be obtained from manufacturers							
Eggshell/semi-gloss finish (oil-based)	11-14	9-11	11-14	8-10	7-9	-	10-13	10-12
Emulsion paint								
standard	12-15	8-12	11-14	8-12	6-10	2-4	12-15	8-10
contract	10-12	7-11	10-12	7-10	5-9	2-4	10-12	7-9
Glossy emulsion	Figures should be obtained from manufacturers							
Heavy textured coating	2-4	2-4	2-4	2-4	2-4	-	2-4	2-4

Surfaces	A	B	C	D	E	F	G	H
Heavy textured coating	2-4	2-4	2-4	2-4	2-4	-	2-4	2-4
Masonry paint per kilogram	5-7	4-6	5-7	4-6	3-5	2-4	-	-
Oil-bound water paint	7-9	4-6	7-9	4-6	5-7	-	-	4-6
Cement-based paint	-	4-6	6-7	3-6	3-6	2-3	-	-
Wood primer (oil-based)	-	-	-	-	-	8-11	-	-

Surfaces	I	J	K	L	M	N	O	P
Water-thinned primer/undercoat								
primer	-	8-11	7-10	-	-	10-14	-	-
undercoat	-	10-12	-	-	-	12-15	12-15	-
Aluminium sealer*								
spirit-based	-	-	-	-	-	7-9	-	-
oil-based	-	-	-	9-13	9-13	-	-	-
Metal primer								
conventional	-	-	-	7-10	10-12	-	-	-
specialised	Figures should be obtained from manufacturers							
Plaster primer (including building board)	8-10	10-12	10-12	-	-	-	-	-

Surfaces	I	J	K	L	M	N	O	P
Alkali-resistant primer	-	-	8-10	-	-	-	-	-
External wall primer sealer	-	6-8	-	-	-	-	-	-
Undercoat	10-12	11-14	10-12	10-12	10-12	10-12	11-14	-
Gloss finish	10-12	11-14	10-12	10-12	10-12	10-12	11-14	11-14
Oil-based thixotropic finish	Figures should be obtained from manufacturers							
Eggshell/semigloss finish (oil-based)	10-12	11-14	10-12	10-12	10-12	10-12	11-14	11-14
Emulsion paint standard	8-10	12-15	10-12	-	-	10-12	12-15	12-15
contract	-	10-12	8-10	-	-	10-12	10-12	10-12
Glossy emulsion	Figures should be obtained from manufacturers							
Heavy textured coating	2-4	2-4	2-4	2-4	2-4	2-4	2-4	2-4
Masonry paint								
per kilogram	-	-	5-7	-	-	-	6-8	6-8
oil-bound water paint	-	7-9	7-9	-	-	-	7-9	-
Cement-based paint	-	-	4-6	-	-	-	-	-

Key

A - Finishing plaster
B - Wood-floated rendering
C - Smooth concrete/cement
D - Fair-faced brickwork
E - Blockvork
F - Roughcast/pebble dash
G - Hardboard
H - Soft fibre insulating board
I - Fire-retardant fibre insulating board
J - Smooth paper-faced board
K - Hard asbestos sheet
L - Structural steelwork
M - Metal sheeting
N - Joinery
O - Smooth primed surfaces
P - Smooth undercoated surfaces

* Aluminium primer/sealer is normally used over bitumen painted surfaces.

In many instances the coverages achieved will be affected by the suction and texture of the backing; for example, the suction and texture of brickwork can vary to such an extent that coverages outside those quoted may, on occasions, be obtained. It is necessary to take these factors into account when using the table.

Labour grades

Craftsman   PA

	Unit	Labour grade	Labour hours	Emulsion litres
One coat matt emulsion paint, surfaces over 300mm girth				
Brickwork				
walls	m2	LA	0.14	0.10
walls in staircase areas	m2	LA	0.16	0.10
Concrete				
walls	m2	LA	0.11	0.08
walls in staircase areas	m2	LA	0.14	0.08
ceilings	m2	LA	0.15	0.08
ceilings in staircase areas	m2	LA	0.17	0.08
Plaster				
walls	m2	LA	0.10	0.07
walls in staircase areas	m2	LA	0.12	0.07
ceilings	m2	LA	0.14	0.07
ceilings in staircase areas	m2	LA	0.15	0.07
Embossed paper				
walls	m2	LA	0.11	0.08
walls in staircase areas	m2	LA	0.14	0.08
ceilings	m2	LA	0.15	0.08
ceilings in staircase areas	m2	LA	0.17	0.08
Two coats matt emulsion paint, surfaces over 300mm girth				
Brickwork				
walls	m2	LA	0.30	0.20
walls in staircase areas	m2	LA	0.34	0.20

	Unit	Labour grade	Labour hours	Emulsion litres
Concrete				
walls	m2	LA	0.24	0.15
walls in staircase areas	m2	LA	0.30	0.15
ceilings	m2	LA	0.32	0.15
ceilings in staircase areas	m2	LA	0.36	0.15
Plaster				
walls	m2	LA	0.23	0.13
walls in staircase areas	m2	LA	0.36	0.13
ceilings	m2	LA	0.30	0.13
ceilings in staircase areas	m2	LA	0.32	0.13
Embossed paper				
walls	m2	LA	0.24	0.15
walls in staircase areas	m2	LA	0.30	0.15
ceilings	m2	LA	0.32	0.15
ceilings in staircase areas	m2	LA	0.36	0.15

	Unit	Labour grade	Labour hours	Primer litres	U/coat litres	Gloss litres
One coat primer sealer, one oil-based undercoat, one coat gloss						
Brickwork						
walls	m2	LA	0.60	0.10	0.16	0.16
walls in staircase areas	m2	LA	0.64	0.10	0.16	0.16

	Unit	Labour grade	Labour hours	Primer litres	U/coat litres	Gloss litres
Concrete						
walls	m2	LA	0.55	0.09	0.13	0.13
walls in staircase						
areas	m2	LA	0.61	0.09	0.13	0.13
ceilings	m2	LA	0.63	0.09	0.13	0.13
ceilings in staircase						
areas	m2	LA	0.67	0.09	0.13	0.13
Plaster						
walls	m2	LA	0.53	0.08	0.08	0.08
walls in staircase						
areas	m2	LA	0.58	0.08	0.08	0.08
ceilings	m2	LA	0.61	0.08	0.08	0.08
ceilings in staircase						
areas	m2	LA	0.64	0.08	0.08	0.08
Embossed paper						
walls	m2	LA	0.55	0.09	0.13	0.13
walls in staircase						
areas	m2	LA	0.61	0.09	0.13	0.13
ceilings	m2	LA	0.63	0.09	0.13	0.13
ceilings in staircase						
areas	m2	LA	0.67	0.09	0.13	0.13

One coat wood primer, one oil-based undercoat, one coat eggshell finish

General surfaces

	Unit	Labour grade	Labour hours	Primer litres	U/coat litres	Gloss litres
over 300mm girth	m2	LA	0.72	0.08	0.09	0.09
isolated surfaces not exceeding 300mm girth	m	LA	0.26	0.03	0.03	0.03
isolated areas not exceeding 0.5m2	nr	LA	0.36	0.04	0.05	0.05

	Unit	Labour grade	Labour hours	Primer litres	U/coat litres	Gloss litres
Windows, screens, glazed doors and the like						
panes under 0.1m2	m2	LA	1.66	0.07	0.08	0.08
panes 0.1-0.5m2	m2	LA	1.43	0.06	0.06	0.06
panes 0.5-1m2	m2	LA	1.23	0.04	0.04	0.04
panes over1m2	m2	LA	1.01	0.02	0.02	0.02

One coat red lead primer, one oil-based undercoat, one coat gloss finish

	Unit	Labour grade	Labour hours	Primer litres	U/coat litres	Gloss litres
General surfaces						
over 300mm girth	m2	LA	0.69	0.08	0.09	0.09
isolated surfaces not exceeding 300mm girth	m	LA	0.26	0.03	0.03	0.03
isolated areas not exceeding 0.5m2	nr	LA	0.36	0.04	0.05	0.05
General surfaces						
over 300mm girth	m2	LA	0.69	0.08	0.09	0.09
isolated surfaces not exceeding 300mm girth	m	LA	0.26	0.03	0.03	0.03
isolated areas not exceeding 0.5m2	nr	LA	0.36	0.04	0.05	0.05
Windows, screens, glazed doors and the like						
panes 0.1m2	m2	LA	1.66	0.07	0.08	0.08
panes 0.1-0.5m2	m2	LA	1.43	0.06	0.06	0.06
panes 0.5-1m2	m2	LA	1.23	0.04	0.04	0.04
panes over 1m2	m2	LA	1.01	0.02	0.02	0.02

	Unit	Labour grade	Labour hours	Primer litres	U/coat litres	Gloss litres
Structural metalwork, general surfaces						
over 300mm girth	m2	LA	0.79	0.08	0.09	0.09
isolated surfaces not exceeding 300mm girth	m	LA	0.29	0.03	0.03	0.03
isolated areas not exceeding 0.5m2	nr	LA	0.35	0.04	0.05	0.05
Structural metalwork, members of trusses, lattice girders, purlins and the like						
over 300mm girth	m2	LA	1.02	0.08	0.09	0.09
isolated surfaces not exceeding 300mm girth	m	LA	0.37	0.03	0.03	0.03
isolated areas not exceeding 0.5m2	nr	LA	0.51	0.04	0.05	0.05
Radiators, panel type						
over 300mm girth	m2	LA	0.98	0.08	0.09	0.09
isolated surfaces not exceeding 300mm girth	m	LA	0.31	0.03	0.03	0.03
isolated areas not exceeding 0.5m2	nr	LA	0.44	0.04	0.05	0.05
Radiators, column type						
over 300mm girth	m2	LA	1.03	0.08	0.09	0.09
isolated surfaces not exceeding 300mm girth	m	LA	0.38	0.03	0.03	0.03
isolated areas not exceeding 0.5m2	nr	LA	0.52	0.04	0.05	0.05

	Unit	Labour grade	Labour hours	Polyurethane litres
Prepare,apply two coats polyurethane, wood				
General surfaces				
over 300mm girth	m2	LA	0.36	0.10
isolated surfaces not exceeding 300mm girth	m	LA	0.10	0.03
isolated areas not exceeding 0.5m2	nr	LA	0.24	0.05
Windows, screens, glazed doors and the like				
panes under 0.1m2	m2	LA	0.80	0.07
panes 0.1-0.5m2	m2	LA	0.72	0.06
panes 0.5-1m2	m2	LA	0.64	0.04
panes over 1m2	m2	LA	0.56	0.02

	Unit	Labour grade	Labour hours	Prep. kg	Sealer litres	Finish litres
One coat Artex preparation, one coat Artex sealer,one coat Artex XL,surfaces over 300mm girth						
Brickwork						
walls	m2	LA	0.48	3.10	0.03	0.50
walls in staircase areas	m2	LA	0.50	3.10	0.03	0.50

	Unit	Labour grade	Labour hours	Prep. kg	Sealer litres	Finish litres
Brickwork						
walls	m2	LA	0.48	3.10	0.03	0.50
walls in staircase areas	m2	LA	0.50	3.10	0.03	0.50
Brickwork						
walls	m2	LA	0.48	2.80	0.03	0.50
walls in staircase	m2	LA	0.50	3.10	0.03	0.50
areas	m2	LA	0.50	3.10	0.03	0.50
Concrete						
walls	m2	LA	0.46	3.80	0.03	0.50
walls in staircase areas	m2	LA	0.48	3.80	0.03	0.50
ceilings	m2	LA	0.52	3.80	0.03	0.50
ceilings in staircase areas	m2	LA	0.54	3.80	0.03	0.50
Plaster						
walls	m2	LA	0.40	2.50	0.03	0.50
walls in staircase areas	m2	LA	0.42	2.50	0.03	0.50
ceilings	m2	LA	0.44	2.50	0.03	0.50
ceilings in staircase areas	m2	LA	0.46	2.50	0.03	0.50
Cement render						
walls	m2	LA	0.42	3.10	0.03	0.50
walls in staircase areas	m2	LA	0.44	3.10	0.03	0.50
ceilings	m2	LA	0.47	3.10	0.03	0.50
ceilings in staircase areas	m2	LA	0.49	3.10	0.03	0.50

	Unit	Labour grade	Labour hours	Creosote litres
One coat creosote to general wrought surfaces				
over 300mm girth	m2	LA	0.20	0.14
isolated surfaces not exceeding 300mm girth	m	LA	0.10	0.04
isolated areas not exceeding 0.5m2	n r	LA	0.14	0.05
One coat creosote to general sawn surfaces				
over 300mm girth	m2	LA	0.22	0.17
isolated surfaces not exceeding 300mm girth	m	LA	0.11	0.05
isolated areas not exceeding 0.5m2	n r	LA	0.15	0.06
Two coats creosote to general wrought surfaces				
over 300mm girth	m2	LA	0.36	0.17
isolated surfaccs not exceeding 300mm girth	m	LA	0.18	0.06
isolated areas not exceeding 0.5m2	n r	LA	0.22	0.08
Two coats creosote to general sawn surfaces				
over 300mm girth	m2	LA	0.40	0.20
isolated surfaces not exceeding 300mm girth	m	LA	0.20	0.07
isolated areas not exceeding 0.5m2	n r	LA	0.26	0.09

	Unit	Labour grade	Labour hours	Polyurethane litres
One coat polyurethane to general wrought surfaces				
over 300mm girth	m2	LA	0.20	0.10
isolated surfaces not exceeding 300mm girth	m	LA	0.09	0.04
isolated areas not exceeding 0.5m2	nr	LA	0.12	0.05
Two coats polyurethane to general wrought surfaces				
over 300mm girth	m2	LA	0.34	0.10
isolated surfaces not exceeding 300mm girth	m	LA	0.16	0.04
isolated areas not exceeding 0.5m2	nr	LA	0.18	0.05

	Unit	Labour grade	Labour hours	Paint litres
One coat masonry stabilising solution to external walls				
Brickwork	m2	LA	0.16	0.16
Blockwork	m2	LA	0.18	0.18
Concrete	m2	LA	0.15	0.16
Cement render	m2	LA	0.14	0.17
Roughcast	m2	LA	0.19	0.17

	Unit	Labour grade	Labour hours	Paint litres
One coat masonry paint to external walls				
Brickwork	m2	LA	0.20	0.16
Blockwork	m2	LA	0.22	0.18
Concrete	m2	LA	0.19	0.16
Cement render	m2	LA	0.18	0.17
Roughcast	m2	LA	0.23	0.17

16

External works

Weights of materials   **kg/m3**

Ashes	800
Bricks, common	1760
engineering	1760
Cement	1900
Clay, dry	1800
Concrete	2300
Gravel	1750
Limestone, crushed	1760
Sand	1600
Hardcore	1900
Topsoil	1000

Blocks/slabs per m2   **Size**   **nr/m2**

Size	nr/m2
200 x 100mm	50.00
450 x 450mm	4.93
600 x 450mm	3.70
600 x 600mm	2.79
600 x 750mm	2.22
600 x 900mm	1.85

Labour grades

Craftsman	LA
Semi-skilled operative	LB
Unskilled operative	LC

Plant grades

Vibrating roller	PE

	Unit	Labour grade	Labour hours	Plant grade	Plant hours	Filling tonnes

Sub-bases

Filling, deposited in
layers 250mm thick,
graded and compacted

	Unit	Labour grade	Labour hours	Plant grade	Plant hours	Filling tonnes
granular fill	m3	LC	0.33	PE	0.10	1.90
sand	m3	LC	0.42	PE	0.10	1.60
hardcore	m3	LC	0.42	PE	0.10	1.90

100mm bed, graded
and compacted

	Unit	Labour grade	Labour hours	Plant grade	Plant hours	Filling tonnes
granular fill	m2	LC	0.05	PE	0.01	0.19
sand	m2	LC	0.05	PE	0.01	0.16
hardcore	m2	LC	0.05	PE	0.01	0.19

150mm bed, graded
and compacted

	Unit	Labour grade	Labour hours	Plant grade	Plant hours	Filling tonnes
granular fill	m2	LC	0.07	PE	0.02	0.27
sand	m2	LC	0.07	PE	0.02	0.29
hardcore	m2	LC	0.07	PE	0.02	0.27

200mm bed, graded
and compacted

	Unit	Labour grade	Labour hours	Plant grade	Plant hours	Filling tonnes
granular fill	m2	LC	0.10	PE	0.03	0.38
sand	m2	LC	0.10	PE	0.03	0.32
hardcore	m2	LC	0.10	PE	0.03	0.38

250mm bed, graded
and compacted

	Unit	Labour grade	Labour hours	Plant grade	Plant hours	Filling tonnes
granular fill	m2	LC	0.12	PE	0.04	0.47
sand	m2	LC	0.12	PE	0.04	0.40
hardcore	m2	LC	0.12	PE	0.04	0.47

	Unit	Labour grade	Labour hours	Plant grade	Plant hours	Filling tonnes
300mm bed, graded and compacted						
granular fill	m2	LC	0.14	PE	0.05	0.57
sand	m2	LC	0.14	PE	0.05	0.48
hardcore	m2	LC	0.14	PE	0.05	0.57

	Unit	Labour grade	Labour hours	Materials tonnes
In situ concrete beds				
Site mixed concrete 1:2:4 (21.00 N/mm2) 20mm aggregate				
beds not exceeding 150mm thick	m3	LB	2.80	2.45
beds 150-450mm thick	m3	LB	1.70	2.45
Formwork to sides of concrete beds				
not exceeding 250mm	m	LA	0.60	-
250-500mm wide	m	LA	0.90	-
Steel fabric reinforcement laid in concrete beds				
A142	m2	LB	0.12	2.22
A193	m2	LB	0.15	3.02
Extra for placing around reinforcement	m2	LB	0.55	-
Expansion joint, impregnated fibre based joint filler, formed joint 12.5mm thick				
not exceeding 150mm wide	m	LB	0.12	-
150-300mm wide	m	LB	0.18	-

	Unit	Labour grade	Labour hours	Materials tonnes
Prepare level surfaces of unset concrete				
mechanical tamping	m2	LB	0.06	-
power floating	m2	LB	0.15	-
trowelling	m2	LB	0.15	-

Gravel pavings

	Unit	Labour grade	Labour hours	Materials tonnes
20mm gravel in bed 60mm thick on bed of clinker aggregate 50mm thick	m2	LB	0.11	0.193
40mm gravel in bed 70mm thick on bed of clinker aggregate 75mm thick	m2	LB	0.15	0.254

	Unit	Labour grade	Labour hours	Bricks nr	Mortar m3
Brick pavings					
Brick paviours, 215 x 103 x 65mm, laid to falls and crossfalls, jointed in cement mortar					
straight joints laid flat	m2	LA	1.00	45.15	0.017
straight joints laid on edge	m2	LA	1.25	73.80	0.025
herringbone pattern laid flat	m2	LA	1.25	45.15	0.017
herringbone pattern laid on edge	m2	LA	1.50	73.80	0.025

	Unit	Labour grade	Labour hours	Flags nr	Mortar m3

Precast concrete pavings

Precast concrete paving
flags spot bedded in
cement mortar

natural/coloured size

	Unit	Labour grade	Labour hours	Flags nr	Mortar m3
450 x 450 x 50mm	m2	LA	0.43	4.93	0.009
600 x 450 x 50mm	m2	LA	0.40	3.70	0.007
600 x 600 x 50mm	m2	LA	0.34	2.79	0.005
600 x 750 x 50mm	m2	LA	0.30	2.22	0.004
600 x 900 x 50mm	m2	LA	0.28	1.85	0.003
Raking cutting on precast concrete flags 50mm thick	m	LA	0.08	-	-
Curved cutting on precast concrete flags 50mm thick	m	LA	0.10	-	-

	Unit	Labour grade	Labour hours	Concrete m3	Mortar m3

**Precast concrete edgings,
kerbs and channels**

Precast concrete edging
kerbs size 50 x 150mm
jointed and pointed in
cement mortar, one side
haunched with concrete

	Unit	Labour grade	Labour hours	Concrete m3	Mortar m3
straight	m	LA	0.40	0.015	0.003
curved to radius not exceeding 3m	m	LA	0.55	0.015	0.003
curved to radius 3-6m	m	LA	0.50	0.015	0.003
curved to radius 6-9m	m	LA	0.45	0.015	0.003

	Unit	Labour grade	Labour hours	Concrete m3	Mortar m3
Precast concrete kerbs size 125 x 150mm jointed and pointed in cement mortar both sides haunched with concrete					
straight	m	LA	0.45	0.015	0.001
curved to radius not exceeding 3m	m	LA	0.60	0.015	0.001
curved to radius 3-6m	m	LA	0.55	0.015	0.001
curved to radius 6-9m	m	LA	0.50	0.015	0.001
Precast concrete channels size 125 x 150mm jointed and pointed in cement mortar, one side haunched with concrete					
straight	m	LA	0.45	0.015	0.001
curved to radius not exceeding 3m	m	LA	0.60	0.015	0.001
curved to radius 3-6m	m	LA	0.55	0.015	0.001
curved to radius 6-9m	m	LA	0.50	0.015	0.001

	Unit	Labour grade	Labour hours	Rails m	Posts nr	Concrete m3
Fencing						
Chestnut fencing						
Chestnut fencing with pales at 75mm centres on two lines of wire fixed to 75mm posts at 3m centres, driven into ground						
900mm high	m	LB	0.45	-	0.33	-
1100mm high	m	LB	0.50	-	0.33	-

	Unit	Labour grade	Labour hours	Rails m	Posts nr	Concrete m3
Chestnut fencing with pales at 75mm centres on three lines of wire fixed to 75mm posts at 3m centres, driven into ground						
900mm high	m	LB	0.47	-	0.33	-
1100mm high	m	LB	0.52	-	0.33	-

Chainlink fencing

Chainlink fencing with line wires fixed to precast concrete posts at 3m centres set in concrete, height

	Unit	Labour grade	Labour hours	Rails m	Posts nr	Concrete m3
900mm	m	LB	0.50	0.90	0.30	0.01
1200mm	m	LB	0.55	1.20	0.30	0.01
1400mm	m	LB	0.60	1.40	0.30	0.01
1800mm	m	LB	0.70	1.80	0.30	0.01
2100mm	m	LB	0.80	1.80	0.30	0.01
2400mm	m	LB	0.90	2.40	0.30	0.01

Post and rail fencing

Post and rail fencing, 100mm half round treated softwood posts at 2m centres set in concrete, two 75mm half round rails, height

	Unit	Labour grade	Labour hours	Rails m	Posts nr	Concrete m3
1200mm	m	LB	0.70	2.00	0.50	0.02
1500mm	m	LB	0.80	2.00	0.50	0.02

	Unit	Labour grade	Labour hours	Panels nr	Posts nr	Concrete m3

Timber panel fencing

Overlapping panel fencing
fixed in slots of concrete
posts at 2m centres set in
concrete, height

900mm	m	LB	0.45	1.00	0.50	0.02
1200mm	m	LB	0.50	1.00	0.50	0.02
1500mm	m	LB	0.55	1.00	0.50	0.02
1800mm	m	LB	0.60	1.00	0.50	0.02

Close boarded panel fencing
fixed in slots of concrete
posts at 2m centres set in
concrete, height

900mm	m	LB	0.45	1.00	0.50	0.02
1200mm	m	LB	0.50	1.00	0.50	0.02
1500mm	m	LB	0.55	1.00	0.50	0.02
1800mm	m	LB	0.60	1.00	0.50	0.02

	Unit	Labour grade	Labour hours	Wire m	Posts nr

Post and wire fencing

100 x 100mm treated
softwood posts driven
into ground at 2m centres

height, 1350mm

3 strands of wire	m	LB	0.35	3.00	0.50
3 strands of single barbed wire	m	LB	0.40	3.00	0.50
3 strands of double barbed wire	m	LB	0.40	3.00	0.50

	Unit	Labour grade	Labour hours	Wire m	Posts nr
height, 1650mm					
4 strands of wire	m	LB	0.40	4.00	0.50
4 strands of single barbed wire	m	LB	0.45	4.00	0.50
4 strands of double barbed wire	m	LB	0.45	4.00	0.50
height, 1950mm					
5 strands of wire	m	LB	0.45	5.00	0.50
5 strands of single barbed wire	m	LB	0.50	5.00	0.50
5 strands of double barbed wire	m	LB	0.50	5.00	0.50

Hurdle fencing

Panel fencing consisting
of osier interwoven hurdles
fixed to treated softwood
stakes driven into ground at
1800mm centres

920mm	m	LB	0.50	-	0.55
1220mm	m	LB	0.55	-	0.55
1520mm	m	LB	0.60	-	0.55
1830mm	m	LB	0.65	-	0.55

	Unit	Labour grade	Labour hours	Wire m	Posts nr

Palisade fencing

Palisade fencing consisting
of 100 x 75mm softwood
posts set in concreted
post holes, two 90 x 38mm
treated softwood rails
and 75 x 19mm treated
softwood pales with pointed
tops at 150mm centres,
overall height

	Unit	Labour grade	Labour hours	Wire m	Posts nr
1000mm	m	LB	0.90	-	0.55
1200mm	m	LB	1.00	-	0.55

17

Drainage

Weights of materials   **kg/m3**

Ashes		800
Bricks, common		1760
engineering		1760
Cement		1900
Concrete		2300
Gravel		1750
Limestone, crushed		1760
Sand		1600

		kg/m
PVC-U pipes	80mm	1.20
	110mm	1.60
	160mm	3.00
	200mm	4.60
	250mm	7.20
Vitrified clay pipes	100mm	15.63
	150mm	37.04
	225mm	95.24
	300mm	196.08
	400mm	357.14
	450mm	500.00
	500mm	555.60

Volumes of filling (m3 per linear metre)

Pipe dia. mm	Beds 50mm	Beds 100mm	Beds 150mm	Bed and haunching	Surround
100	0.023	0.045	0.068	0.117	0.185
150	0.026	0.053	0.079	0.152	0.231
225	0.030	0.060	0.090	0.195	0.285
300	0.038	0.075	0.113	0.279	0.391

Trench widths

Pipe dia. mm	Less than 1.5m deep mm	More than 1.5m deep mm
100	450	600
150	500	650
225	600	750
300	650	800
400	750	900
450	900	1050
600	1000	1300

Labour grades

Semi-skilled operative	LB
Unskilled operative	LC
2 Bricklayers and 1 unskilled operative	LD

Plant grades

Hydraulic excavator (1.7m3)	PA

	Unit	Labour grade	Labour hours	Trench m3

Hand excavation

Excavate trench for drain,
grade and ram bottom,
backfill with excavated
material, for pipe diameter
100mm, average depth

	Unit	Labour grade	Labour hours	Trench m3
0.50m	m	LB	1.20	0.23
0.75m	m	LB	2.00	0.23
1.00m	m	LB	2.50	0.23
1.25m	m	LB	4.00	0.23
1.50m	m	LB	5.10	0.23
1.75m	m	LB	5.90	1.05
2.00m	m	LB	6.30	1.20
2.25m	m	LB	8.25	1.35
2.50m	m	LB	9.45	1.50
2.75m	m	LB	10.75	1.65
3.00m	m	LB	12.00	1.80

Excavate trench for drain,
grade and ram bottom,
backfill with excavated
material, for pipe diameter
150mm, average depth

	Unit	Labour grade	Labour hours	Trench m3
0.50m	m	LB	1.30	0.25
0.75m	m	LB	2.10	0.25
1.00m	m	LB	2.80	0.25
1.25m	m	LB	4.50	0.25
1.50m	m	LB	5.60	0.25
1.75m	m	LB	6.50	1.14
2.00m	m	LB	7.00	1.30
2.25m	m	LB	9.00	1.46
2.50m	m	LB	10.40	1.63
2.75m	m	LB	11.80	1.79
3.00m	m	LB	13.20	1.95

	Unit	Labour grade	Labour hours	Trench m3
Excavate trench for drain, grade and ram bottom, backfill with excavated material, for pipe diameter 225mm, average depth				
0.50m	m	LB	1.40	0.30
0.75m	m	LB	2.20	0.30
1.00m	m	LB	2.90	0.30
1.25m	m	LB	4.70	0.30
1.50m	m	LB	5.80	0.30
1.75m	m	LB	6.75	1.32
2.00m	m	LB	7.40	1.50
2.25m	m	LB	9.45	1.62
2.50m	m	LB	10.85	1.75
2.75m	m	LB	12.40	2.00
3.00m	m	LB	13.85	2.25
Excavate trench for drain, grade and ram bottom, backfill with excavated material, for pipe diameter 300mm, average depth				
0.50m	m	LB	1.50	0.33
0.75m	m	LB	2.30	0.33
1.00m	m	LB	3.00	0.33
1.25m	m	LB	4.90	0.33
1.50m	m	LB	6.00	0.33
1.75m	m	LB	7.00	1.40
2.00m	m	LB	7.60	1.60
2.25m	m	LB	9.90	1.80
2.50m	m	LB	11.50	2.00
2.75m	m	LB	13.00	2.20
3.00m	m	LB	14.50	2.40

	Unit	Labour grade	Labour hours	Plant grade	Plant hours	Trench m3

Machine excavation

Excavate trench for drain,
grade and ram bottom,
backfill with excavated
material, for pipe diameter
less than 100mm, average
depth

0.50m	m	LB	0.20	PA	0.10	0.23
0.75m	m	LB	0.30	PA	0.15	0.23
1.00m	m	LB	0.60	PA	0.45	0.23
1.25m	m	LB	0.90	PA	0.55	0.23
1.50m	m	LB	1.15	PA	0.70	0.23
1.75m	m	LB	1.40	PA	0.80	1.05
2.00m	m	LB	1.65	PA	0.90	1.20
2.25m	m	LB	1.80	PA	0.95	1.35
2.50m	m	LB	1.90	PA	1.00	1.50
2.75m	m	LB	2.00	PA	1.15	1.65
3.00m	m	LB	2.35	PA	1.25	1.80

Excavate trench for drain,
grade and ram bottom,
backfill with excavated
material, for pipe diameter
150mm, average depth

0.50m	m	LB	0.22	PA	0.10	0.25
0.75m	m	LB	0.32	PA	0.15	0.25
1.00m	m	LB	0.62	PA	0.35	0.25
1.25m	m	LB	0.92	PA	0.55	0.25
1.50m	m	LB	1.18	PA	0.70	0.25
1.75m	m	LB	1.43	PA	0.80	1.14
2.00m	m	LB	1.68	PA	0.90	1.30
2.25m	m	LB	1.84	PA	0.95	1.46
2.50m	m	LB	1.94	PA	1.00	1.63
2.75m	m	LB	2.05	PA	1.15	1.79
3.00m	m	LB	2.40	PA	1.25	1.95

	Unit	Labour grade	Labour hours	Plant grade	Plant hours	Trench m3

Excavate trench for drain,
grade and ram bottom,
backfill with excavated
material, for pipe diameter
225mm, average depth

0.50m	m	LB	0.24	PA	0.12	0.30
0.75m	m	LB	0.36	PA	0.17	0.30
1.00m	m	LB	0.72	PA	0.38	0.30
1.25m	m	LB	1.08	PA	0.60	0.30
1.50m	m	LB	1.38	PA	0.75	0.30
1.75m	m	LB	1.68	PA	0.85	1.32
2.00m	m	LB	1.98	PA	0.95	1.50
2.25m	m	LB	2.16	PA	1.05	1.62
2.50m	m	LB	2.28	PA	1.10	1.75
2.75m	m	LB	2.40	PA	1.20	2.00
3.00m	m	LB	2.80	PA	1.35	2.25

Excavate trench for drain,
grade and ram bottom,
backfill with excavated
material, for pipe diameter
300mm, average depth

0.50m	m	LB	0.26	PA	0.12	0.33
0.75m	m	LB	0.40	PA	0.17	0.33
1.00m	m	LB	0.82	PA	0.38	0.33
1.25m	m	LB	1.24	PA	0.60	0.33
1.50m	m	LB	1.58	PA	0.75	0.33
1.75m	m	LB	1.93	PA	0.85	1.40
2.00m	m	LB	2.18	PA	0.95	1.60
2.25m	m	LB	2.48	PA	1.05	1.80
2.50m	m	LB	2.60	PA	1.10	2.00
2.75m	m	LB	2.75	PA	1.20	2.30
3.00m	m	LB	3.20	PA	1.35	2.40

	Unit	Labour grade	Labour hours	Materials m3
Beds and coverings				
Sand bed 100mm thick under pipe diameter				
100mm	m	LB	0.10	0.045
150mm	m	LB	0.11	0.052
225mm	m	LB	0.14	0.060
Sand bed 150mm thick under pipe diameter				
100mm	m	LB	0.11	0.068
150mm	m	LB	0.13	0.079
225mm	m	LB	0.15	0.090
Granular bed 100mm thick under pipe diameter				
100mm	m	LB	0.12	0.045
150mm	m	LB	0.13	0.052
225mm	m	LB	0.16	0.060
Granular bed 150mm thick under pipe diameter				
100mm	m	LB	0.14	0.068
150mm	m	LB	0.15	0.079
225mm	m	LB	0.18	0.090
Concrete bed 100mm thick under pipe diameter				
100mm	m	LB	0.24	0.045
150mm	m	LB	0.26	0.052
225mm	m	LB	0.32	0.060

	Unit	Labour grade	Labour hours	Materials m3
Concrete bed 150mm thick under pipe diameter				
100mm	m	LB	0.28	0.068
150mm	m	LB	0.30	0.079
225mm	m	LB	0.36	0.090
Granular bed and haunching to pipe diameter				
100mm	m	LB	0.24	0.117
150mm	m	LB	0.26	0.152
225mm	m	LB	0.32	0.195
Concrete bed and haunching to pipe diameter				
100mm	m	LB	0.48	0.117
150mm	m	LB	0.52	0.152
225mm	m	LB	0.64	0.195
Granular bed and surround topipe diameter				
100mm	m	LB	0.36	0.185
150mm	m	LB	0.39	0.231
225mm	m	LB	0.48	0.285
Concrete bed and surround to pipe diameter				
100mm	m	LB	0.72	0.185
150mm	m	LB	0.78	0.231
225mm	m	LB	0.90	0.285

	Unit	Labour grade	Labour hours	Gaskin m	Mortar m3	Coupling nr
Pipework						
Vitrified clay drain pipe with push-fit joints laid in trenches, 100mm pipe diameter						
Laid straight	m	LB	0.20	-	-	0.66
Less than 3m runs	m	LB	0.25	-	-	0.66
Bends	nr	LB	0.20	-	-	2.00
Rest bends	nr	LB	0.20	-	-	2.00
Junctions	nr	LB	0.20	-	-	2.00
Adaptor	nr	LB	0.25	-	-	2.00
Vitrified clay drain pipe with push-fit joints laid in trenches, 150mm pipe diameter						
Laid straight	m	LB	0.25	-	-	0.66
Less than 3m runs	m	LB	0.30	-	-	0.66
Bends	nr	LB	0.25	-	-	2.00
Rest bends	nr	LB	0.25	-	-	2.00
Junctions	nr	LB	0.25	-	-	2.00
Adaptor	nr	LB	0.30	-	-	2.00
Vitrified clay drain pipe with push-fit joints laid in trenches, 225mm pipe diameter						
Laid straight	m	LB	0.30	-	-	0.66
Less than 3m runs	m	LB	0.35	-	-	0.66
Bends	nr	LB	0.30	-	-	2.00
Rest bends	nr	LB	0.30	-	-	2.00
Junctions	nr	LB	0.30	-	-	2.00
Adaptor	nr	LB	0.35	-	-	2.00

	Unit	Labour grade	Labour hours	Gaskin m	Mortar m3	Coupling nr
Vitrified clay drain pipe with push-fit joints laid in trenches, 300mm pipe diameter						
Laid straight	m	LB	0.40	-	-	0.66
Less than 3m runs	m	LB	0.45	-	-	0.66
Bends	nr	LB	0.40	-	-	2.00
Rest bends	nr	LB	0.40	-	-	2.00
Junctions	nr	LB	0.40	-	-	2.00
Adaptor	nr	LB	0.45	-	-	2.00
Vitrified clay drain pipe with spigot and socket joints in gaskin and cement mortar, laid in trenches, 100mm pipe diameter						
Laid straight	m	LB	0.35	0.30	0.015	-
Less than 3m runs	m	LB	0.40	0.30	0.015	-
Bends	nr	LB	0.35	0.60	0.030	-
Rest bends	nr	LB	0.35	0.60	0.030	-
Junctions	nr	LB	0.35	0.60	0.030	-
Adaptor	nr	LB	0.40	0.60	0.030	-
Vitrified clay drain pipe with spigot and socket joints in gaskin and cement mortar, laid in trenches, 150mm pipe diameter						
Laid straight	m	LB	0.45	0.50	0.018	-
Less than 3m runs	m	LB	0.50	0.50	0.018	-
Bends	nr	LB	0.45	1.00	0.036	-
Rest bends	nr	LB	0.45	1.00	0.036	-
Junctions	nr	LB	0.45	1.00	0.036	-
Adaptor	nr	LB	0.50	1.00	0.036	-

	Unit	Labour grade	Labour hours	Gaskin m	Mortar m3	Coupling nr
Vitrified clay drain pipe with spigot and socket joints in gaskin and cement mortar, laid in trenches, 225mm pipe diameter						
Laid straight	m	LB	0.55	0.70	0.020	-
Less than 3m runs	m	LB	0.60	0.70	0.020	-
Bends	nr	LB	0.55	1.40	0.040	-
Rest bends	nr	LB	0.55	1.40	0.040	-
Junctions	nr	LB	0.55	1.40	0.040	-
Adaptor	nr	LB	0.60	1.40	0.040	-
Vitrified clay drain pipe with spigot and socket joints in gaskin and cement mortar, laid in trenches, 300mm pipe diameter						
Laid straight	m	LB	0.80	1.00	0.025	-
Less than 3m runs	m	LB	0.90	1.00	0.025	-
Bends	nr	LB	0.80	2.00	0.050	-
Rest bends	nr	LB	0.80	2.00	0.050	-
Junctions	nr	LB	0.80	2.00	0.050	-
Adaptor	nr	LB	0.90	2.00	0.050	-
PVC-U drain pipe with ring seal joints, laid in trenches, 82mm pipe diameter						
Laid straight	m	LB	0.15	-	-	0.17
Less than 3m runs	m	LB	0.20	-	-	0.17
Bends	nr	LB	0.15	-	-	0.34
Junctions	nr	LB	0.15	-	-	0.34
Adaptor to clay	nr	LB	0.20	-	-	0.34

	Unit	Labour grade	Labour hours	Gaskin m	Mortar m3	Coupling nr
PVC-U drain pipe with ring seal joints, laid in trenches, 110mm pipe diameter						
Laid straight	m	LB	0.18	-	-	0.17
Less than 3m runs	m	LB	0.22	-	-	0.17
Bends	nr	LB	0.18	-	-	0.34
Junctions	nr	LB	0.18	-	-	0.34
Adaptor to clay	nr	LB	0.22	-	-	0.34
PVC-U drain pipe with ring seal joints, laid in trenches, 160mm pipe diameter						
Laid straight	m	LB	0.22	-	-	0.17
Less than 3m runs	m	LB	0.25	-	-	0.17
Bends	nr	LB	0.22	-	-	0.34
Junctions	nr	LB	0.22	-	-	0.34
Adaptor to clay	nr	LB	0.25	-	-	0.34

Accessories

	Unit	Labour grade	Labour hours	Gaskin m	Mortar m3	Coupling nr
Vitrified clay gully with 100mm diameter outlet, 150mm square gulley grid, jointed to drain, surrounded with concrete	nr	LB	1.50	-	-	-
Vitrified clay trapped yard gully with 100mm diameter outlet, 200mm square gulley grid, jointed to drain, surrounded with concrete	nr	LB	2.00	-	-	-

	Unit	Labour grade	Labour hours	Plant grade	Plant hours

Manholes

Excavate for manholes
including earthwork support,
disposal of surplus
excavated material and
compacting bottom of
excavation

By hand, depth not exceeding

	Unit	Labour grade	Labour hours	Plant grade	Plant hours
1.00m	m3	LC	4.00	-	-
2.00m	m3	LC	4.50	-	-
4.00m	m3	LC	5.00	-	-

By machine, depth not exceeding

	Unit	Labour grade	Labour hours	Plant grade	Plant hours
1.00m	m3	LC	0.25	PA	0.25
2.00m	m3	LC	0.30	PA	0.30
4.00m	m3	LC	0.35	PA	0.35

Ready mixed concrete
1:3:6 (11.5N/mm2,
40mm aggregate) in
manhole base, thickness

	Unit	Labour grade	Labour hours	Plant grade	Plant hours
less than 150mm	m3	LC	2.00	-	-
150 to 450mm	m3	LC	1.75	-	-
over 450mm	m3	LC	1.50	-	-

	Unit	Labour grade	Labour hours	Plant grade	Plant hours
Common bricks in cement mortar in one brick thick walls of manholes	m2	LD	3.80	-	-
Engineering bricks in cement mortar in one brick thick walls of manholes	m2	LD	4.00	-	-
Extra for fair face flush pointing	m2	LD	0.20	-	-

	Unit	Labour grade	Labour hours	Plant grade	Plant hours
Best quality vitrified clay channels bedded in cement mortar					
Half round straight main channel					
100mm diameter	m	LD	0.90	-	
150mm diameter	m	LD	1.00	-	-
Half round tapered main channel, 100 to 150mm diameter	m	LD	1.00	-	-
Half round channel bends					
100mm diameter	nr	LD	1.00	-	-
150mm diameter	nr	LD	1.10	-	-
Three quarter section channel bends					
100mm diameter	nr	LD	1.00	-	-
150mm diameter	nr	LD	1.10	-	-
Galvanised step irons built into side of manhole walls	m2	LD	0.10	-	-
Cast iron manhole covers bedded in cement mortar					
Grade C light duty, size 600 x 450mm	nr	LB	1.40	-	
Grade B medium duty, single seal, size 600 x 450mm	nr	LB	1.50	-	

18

Elemental percentage breakdowns

The cost of most buildings can be broken down into about 25 discrete standard sections or elements. In the early planning stages of a project and in the tender analysis process, values are allocated to each element and any imbalances can be identified.

These values can also be expressed as percentages of the overall cost and the following tables display these percentages for 24 different types of buildings. It should be noted that the analyses cover building costs only and exclude contingencies, land values and professional fees. The following types of buildings are covered:

A	Local Authority mixed housing development
B	Flats and maisonettes
C	Sheltered housing
D	Primary school
E	Middle secondary school
F	Sixth form college
G	Advance factory units
H	Factory built for owner-occupation
I	Warehouse (shell only), low bay, 10m high to eaves
J	Two storey office
K	Multi-storey car park
L	Fire station
M	Police station
N	Ambulance station
O	Health centre
P	Welfare centre
Q	Old persons' home
R	Community centre
S	Sports hall
T	Sports pavilion
U	Retail shops
V	Private housing
W	Banks
X	Garage/showrooms

The figures have been rounded off to the nearest whole number and there may be slight distortion in some cases.

	A %	A %	B %	B %	C %	C %	D %	D %
Preliminaries	-	15	-	17	-	9	-	13
Substructure	-	10	-	12	-	9	-	9
Superstructure								
Frame	1		-		-		7	
Upper floors	2		13		4		1	
Roof	7		5		4		10	
Staircases	2		2		1		1	
External walls	9		6		7		10	
Windows and external doors	8		7		6		2	
Internal walls and partitions	7		4		6		3	
Internal doors	4	40	3	40	5	33	2	36
Finishes								
Wall finishes	2		3		5		1	
Floor finishes	4		1		3		2	
Ceiling finishes	3	9	1	5	3	11	2	5
Fittings and furnishings	-	2	-	3	-	3	-	3
Services								
Sanitary appliances and disposal	1		1		2		2	
Services equipment	-		-		2		1	
Heat source	2		3		5		5	
Hot and cold water services	-		-		1		1	
Heating and air treatment	3		6		-		1	
Ventilation installation	-		-		1		-	
Gas services	-		-		-		1	
Electrical installation	3		4		5		5	
Lift and conveyor	-		-		2		-	
Protective communication	-		-		2		-	
Communications installation	-		-		1		1	
Special installation equipment	1		2		1		-	
Builders' work and profit	1	11	-	16	3	25	1	18
External works		13		7		10		16
		100		100		100		100

	%	E %	%	F %	%	G %	%	H %
Preliminaries	-	4	-	12	-	9	-	4
Substructure	-	5	-	8	-	14	-	11
Superstructure								
Frame	6		-		12		12	
Upper floors	1		3		1		1	
Roof	10		13		11		6	
Staircases	1		1		-		6	
External walls	5		7		18		18	
Windows and external doors	3		4		2		1	
Internal walls and partitions	2		3		3		1	
Internal doors	2	30	3	34	2	49	2	47
Finishes								
Wall finishes	2		3		1		4	
Floor finishes	3		4		1		2	
Ceiling finishes	2	7	4	11	1	3	1	7
Fittings and furnishings	-	6	-	3	-	-	-	1
Services								
Sanitary appliances and disposal	2		1		1		2	
Services equipment	1		1		-		-	
Heat source	3		4		1		1	
Hot and cold water services	1		2		1		1	
Heating and air treatment	4		10		-		-	
Ventilation installation	-		-		1		2	
Gas services	-		-		1		-	
Electrical installation	5		7		-		5	
Lift and conveyor	-		-		-		-	
Protective communication	-		-		-		1	
Communications installation	-		-		-		-	
Special installation equipment	-				1		-	
Builders' work and profit	1	17	1	26	-	6	1	13
External works		31		6		19		17
		100		100		100		100

	I %	%	J %	%	K %	%	L %	%
Preliminaries	-	4	-	6	-	9	-	7
Substructure	-	5	-	7	-	14	-	4
Superstructure								
Frame	6		-		12		4	
Upper floors	1		2		1		1	
Roof	10		9		11		4	
Staircases	1		2		-		1	
External walls	5		9		18		7	
Windows and external doors	3		6		2		4	
Internal walls and partitions	2		2		3		4	
Internal doors	2	30	2	32	2	49	3	28
Finishes								
Wall finishes	2		1		1		3	
Floor finishes	3		1		1		4	
Ceiling finishes	2	7	2	4	1	3	1	8
Fittings and furnishings	-	6	-	3	-	-	-	2
Services								
Sanitary appliances and disposal	2		2		1		2	
Services equipment	1		-		-		-	
Heat source	3		1		1		1	
Hot and cold water services	1		2		1		1	
Heating and air treatment	4		3		-		5	
Ventilation installation	-		3		1		3	
Gas services	-		1		1		1	
Electrical installation	5		8		-		10	
Lift and conveyor	-		1		-		-	
Protective communication	-		-		-		1	
Communications installation	-		-		-		1	
Special installation equipment	-		1		1		-	
Builders' work and profit	1	17	-	22	-	6	2	27
External works		31		26		19		24
		100		100		100		100

	M		N		O		P	
	%	%	%	%	%	%	%	%
Preliminaries	-	11	-	6	-	8	-	23
Substructure	-	5	-	7	-	11	-	7
Superstructure								
Frame	5		-		-		-	
Upper floors	2		2		-		-	
Roof	1		9		11		6	
Staircases	5		2		-		-	
External walls	5		9		8		5	
Windows and external doors	7		6		5		8	
Internal walls and partitions	3		2		3		3	
Internal doors	4	32	2	32	5	32	3	25
Finishes								
Wall finishes	2		1		2		2	
Floor finishes	2		1		3		2	
Ceiling finishes	2	6	2	4	3	8	1	5
Fittings and furnishings	-	3	-	3	-	5	-	4
Services								
Sanitary appliances and disposal	2		2		3		2	
Services equipment	1		-		1		-	
Heat source	2		1		3		-	
Hot and cold water services	1		2		3		-	
Heating and air treatment	3		3		3		8	
Ventilation installation	2		3		-		-	
Gas services	1		1		-		-	
Electrical installation	7		8		3		5	
Lift and conveyor	1		1		-		1	
Protective communication	1		-		1		-	
Communications installation	4		-		2		1	
Special installation equipment	-		1		1		-	
Builders' work and profit	1	26	-	22	2	22	1	18
External works		17		26		14		18
		100		100		100		100

	Q		R		S		T
%	%	%	%	%	%	%	%
Preliminaries	-	13	-	26	-	6	-	16
Substructure	-	15	-	7	-	17	-	11

Superstructure

	%	Q %	%	R %	%	S %	%	T %
Frame	1		-		-		-	
Upper floors	1		1		-		-	
Roof	6		9		8		17	
Staircases	1		1		-		-	
External walls	6		8		9		13	
Windows and external doors	6		3		5		1	
Internal walls and partitions	3		2		4		1	
Internal doors	4	28	2	26	3	29	1	33

Finishes

	%	Q %	%	R %	%	S %	%	T %
Wall finishes	4		6		2		1	
Floor finishes	3		2		3		5	
Ceiling finishes	1	8	4	12	3	8	1	7
Fittings and furnishings	-	3	-	2	-	5	-	3

Services

	%	Q %	%	R %	%	S %	%	T %
Sanitary appliances and disposal	3		2		3		2	
Services equipment	1		-		1		1	
Heat source	1		3		3		3	
Hot and cold water services	1		-		3		-	
Heating and air treatment	4		-		3		7	
Ventilation installation	-		-		-		3	
Gas services	1		-		-		1	
Electrical installation	6		2		3		4	
Lift and conveyor	1		-		-		-	
Protective communication	1		-		1		-	
Communications installation	1		-		2		-	
Special installation equipment	-		-		1		-	
Builders' work and profit	1	21	1	8	2	22	2	23
External works		12		19		13		7
		100		100		100		100

		U		V		W		X
	%	%	%	%	%	%	%	%
Preliminaries	-	11	-	18	-	14	-	21
Substructure	-	13	-	11	-	7	-	10
Superstructure								
Frame	4		4		2		5	
Upper floors	-		5		3		-	
Roof	9		5		2		5	
Staircases	-		1		2		-	
External walls	12		7		4		7	
Windows and external doors	8		6		4		4	
Internal walls and partitions	8		5		4		4	
Internal doors	2	45	3	36	2	23	3	28
Finishes								
Wall finishes	1		4		5		2	
Floor finishes	5		2		4		3	
Ceiling finishes	2	8	2	8	3	12	2	7
Fittings and furnishings	-	11	-	-	-	12	-	4
Services								
Sanitary appliances and disposal	1		2		2		2	
Services equipment	-		2		1		1	
Heat source	2		4		2		3	
Hot and cold water services	1		2		2		1	
Heating and air treatment	2		-		1		1	
Ventilation installation	-		-		1		-	
Gas services	-		-		1		1	
Electrical installation	3		4		4		4	
Lift and conveyor	-		-		-		-	
Protective communication	-		-		2		1	
Communications installation	-		-		1		2	
Special installation equipment	1		-		1		6	
Builders' work and profit	1	11	2	16	2	20	2	24
External works		3		9		2		6
		100		100		100		100

PART TWO

CIVIL ENGINEERING

19

Demolition

Labour grades

Ganger and unskilled operative   LE

Plant grades

Hydraulic excavator (1.7m3), crawler
dozer and 6 wheel tipper wagon   PP

	Unit	Labour grade	Labour hours	Plant grade	Plant hours

Demolish buildings to 500mm below ground level, volume

50-100m3

brickwork	nr	LE	6.00	PP	6.00
concrete	nr	LE	7.00	PP	7.00
masonry	nr	LE	6.00	PP	6.00
steel framed	nr	LE	7.00	PP	7.00
timber	nr	LE	4.00	PP	4.00

250-500m3

brickwork	nr	LE	20.00	PP	20.00
concrete	nr	LE	24.00	PP	24.00
masonry	nr	LE	20.00	PP	20.00
steel framed	nr	LE	24.00	PP	24.00
timber	nr	LE	10.00	PP	10.00

1000-2500m3

brickwork	nr	LE	60.00	PP	60.00
concrete	nr	LE	80.00	PP	80.00
masonry	nr	LE	60.00	PP	60.00
steel framed	nr	LE	80.00	PP	80.00
timber	nr	LE	40.00	PP	40.00

Demolish reinforced concrete tanks and the like, volume

not exceeding 50m3	nr	LE	15.00	PP	15.00
50-100m3	nr	LE	21.00	PP	21.00
100-250m3	nr	LE	42.00	PP	42.00
250-500m3	nr	LE	48.00	PP	48.00
500-1000m3	nr	LE	80.00	PP	80.00
1000-2500m3	nr	LE	160.00	PP	160.00

20

Site clearance

Labour grades

Ganger and unskilled operative LE

Plant grades

Hydraulic excavator (1.7m3), crawler
dozer and 6 wheel tipper wagon PP

	Unit	Labour grade	Labour hours	Plant grade	Plant hours
Clear undergrowth, hedges, small trees and vegetation	ha	LE	14.00	PP	14.00
Cut down trees, girth					
0.5-1m	nr	LE	0.50	PP	0.50
1-2m	nr	LE	1.00	PP	1.00
2-3m	nr	LE	3.00	PP	3.00
3-5m	nr	LE	6.00	PP	6.00
Dig up stumps, diameter					
150-500mm	nr	LE	0.50	PP	0.50
500mm-1m	nr	LE	1.00	PP	1.00
2-3m	nr	LE	2.00	PP	3.00
3-5m	nr	LE	3.00	PP	3.00
Dig up and remove existing pipelines, average 1.5m depth					
Clay pipes, nominal bore					
100-300mm	m	LE	0.10	PP	0.10
300-500mm	m	LE	0.12	PP	0.12
Concrete pipes, nominal bore					
100-300mm	m	LE	0.12	PP	0.12
300-500mm	m	LE	0.14	PP	0.14
500-1000mm	m	LE	0.16	PP	0.16
Cast iron pipes, nominal bore					
100-300mm	m	LE	0.14	PP	0.14
300-500mm	m	LE	0.16	PP	0.16

21

Excavation and filling

Weights of materials	kg/m3
Ashes	800
Ballast	600
Chalk	2240
Clay	1800
Flint	2550
Gravel	1750
Hardcore	1900
Hoggin	1750
Lime, ground	750
Sand	1600
Water	950
Shrinkage of deposited materials	**%**
Clay	10.0
Gravel	7.5
Sandy soil	12.5

Bulking of excavated material	%
Clay	+40
Gravel	+25
Sand	+20
Rock, unweathered	+70
Vegetable soil	+30

Angle of repose	Type	Angle degrees
Earth	loose, dry	36-40
	loose, moist	45
	loose, wet	30
	consolidated, dry	42
	consolidated, moist	38
Loam	loose, dry	40-45
	loose,wet	20-25
Gravel	dry	35-45
	wet	25-30
Sand	loose, dry	35-40
	compact	30-35
	wet	25
Clay	loose, wet	20-25
	consolidated, moist	70

Typical fuel consumption for plant

These figures relate to working in normal conditions. Reduce by 25% for light duties and increase by 50% for heavy duties.

Plant	Engine size kW	Litres/ hour
Compressors up to	20	4.0
	30	6.5
	40	8.2
	50	9.0
	75	16.0
	100	20.0
	125	25.0
	150	30.0
Concrete mixers up to	5	1.0
	10	2.4
	15	3.8
	20	5.0
Dumpers	5	1.3
	7	2.0
	10	3.0
	15	4.0
	20	4.9
	30	7.0
	50	12.0
Excavators	10	2.5
	20	4.5
	40	9.0
	60	13.0
	80	17.0
Pumps	5	1.1
	7.5	1.6
	10	2.1
	15	3.2
	20	4.2
	25	5.5

Plant	Engine size kW	Litres/ hour
Trenchers	25	5.0
	35	6.5
	50	10.0
	75	14.5

Average plant outputs (m3/hour)

Bucket size (litres)	Soil	Sand	Heavy clay	Soft rock
Face shovel				
200	11	12	7	5
300	18	20	12	9
400	24	26	17	13
600	42	45	28	23
Backactor				
200	8	8	6	4
300	12	13	9	7
400	17	18	11	10
600	28	30	19	15
Dragline				
200	11	12	8	5
300	18	20	12	9
400	25	27	16	12
600	42	45	28	21

Labour grades

Craftsman	LA
Semi-skilled operative	LB
Unskilled operative	LC
Ganger and unskilled operative	LE

Plant grades

Hydraulic excavator (1.7m3)	PA
Compressor (375cfm)	PB
Skip (8m3)	PC
Tipper wagon (6 wheel)	PD
Vibrating roller	PE
Hydraulic excavator (3.5m3)	PK
Compressor, drills and breakers, hydraulic excavator (3.5m3)	PL
Crawler dozer and grader	PM
Tractor loader and vibrating roller	PO
Hydraulic excavator (1.7m3), crawler dozer and 6 wheel tipper wagon	PP

	Unit	Labour grade	Labour hours	Plant grade	Plant hours
Excavation					
Excavating in normal ground conditions, depth					
0.25-0.5m	m3	LE	0.05	PK	0.05
0.5-1m	m3	LE	0.05	PK	0.05
1-2m	m3	LE	0.06	PK	0.06
2-5m	m3	LE	0.07	PK	0.07
5-10m	m3	LE	0.10	PK	0.10
Excavating in rock at ground level, depth					
0.25-0.5m	m3	LE	0.38	PL	0.38
0.5-1m	m3	LE	0.43	PL	0.43
1-2m	m3	LE	0.54	PL	0.54
2-5m	m3	LE	0.65	PL	0.65
5-10m	m3	LE	0.80	PL	0.80
Excavating in mass concrete at ground level, depth					
0.25-0.5m	m3	LE	0.60	PL	0.60
0.5-1m	m3	LE	0.70	PL	0.60
Excavating in reinforced concrete at ground level, thickness					
0.25-0.5m	m3	LE	0.80	PL	0.60
0.5-1m	m3	LE	0.90	PL	0.60
Excavating in tarmacadam at ground level, thickness					
0.25-0.5m	m3	LE	0.15	PL	0.15

	Unit	Labour grade	Labour hours	Plant grade	Plant hours
Excavating in mass concrete below ground level, thickness					
0.25-0.5m	m3	LE	0.70	PL	0.70
0.5-1m	m3	LE	0.80	PL	0.80
Excavating in reinforced concrete below ground level, thickness					
0.25-0.5m	m3	LE	0.90	PL	0.90
0.5-1m	m3	LE	1.00	PL	1.00

Excavation sundries

Trimming excavated surfaces

	Unit	Labour grade	Labour hours	Plant grade	Plant hours
horizontally	m2	-	-	PM	0.01
10-45° to the horizontal	m2	-	-	PM	0.01
45-90° to the horizontal	m2	-	-	PM	0.02

Trimming subsoil

	Unit	Labour grade	Labour hours	Plant grade	Plant hours
horizontally	m2	-	-	PM	0.01
10-45° to the horizontal	m2	-	-	PM	0.01
45-90° to the horizontal	m2	-	-	PM	0.02

Trimming rock

	Unit	Labour grade	Labour hours	Plant grade	Plant hours
horizontally	m2	-	-	PM	0.35
10-45° to the horizontal	m2	-	-	PM	0.40
45-90° to the horizontal	m2	-	-	PM	0.45

Preparation of excavated surfaces

	Unit	Labour grade	Labour hours	Plant grade	Plant hours
horizontally	m2	-	-	PM	0.02
10-45° to the horizontal	m2	-	-	PM	0.03
45-90° to the horizontal	m2	-	-	PM	0.04

	Unit	Labour grade	Labour hours	Plant grade	Plant hours
Preparation of excavated surfaces					
horizontally	m2	-	-	PM	0.02
10-45° to the horizontal	m2	-	-	PM	0.03
45-90° to the horizontal	m2	-	-	PM	0.04
Preparation of rock					
horizontally	m2	-	-	PM	0.40
10-45° to the horizontal	m2	-	-	PM	0.45
45-90° to the horizontal	m2	-	-	PM	0.50
Disposal					
Disposal of vegetable soil					
remove from site to storage 5km distance	m3	-	-	PN	0.08
remove from site to storage 10km distance	m3	-	-	PN	0.12
remove from site to storage 15km distance	m3	-	-	PN	0.18
remove from site to storage 20km distance	m3	-	-	PN	0.22
store on site 100m distance	m3	-	-	PN	0.05
store on site 200m distance	m3	-	-	PN	0.08

	Unit	Labour grade	Labour hours	Plant grade	Plant hours
Disposal of excavated material					
remove from site to storage 5km distance	m3	-	-	PN	0.08
remove from site to storage 10km distance	m3	-	-	PN	0.12
remove from site to storage 15km distance	m3	-	-	PN	0.18
remove from site to storage 20km distance	m3	-	-	PN	0.22
store on site 100m distance	m3	-	-	PN	0.05
store on site 200m distance	m3	-	-	PN	0.08
Disposal of rock					
remove from site to storage 5km distance	m3	-	-	PN	0.12
remove from site to storage 10km distance	m3	-	-	PN	0.16
remove from site to storage 15km distance	m3	-	-	PN	0.25
remove from site to storage 20km distance	m3	-	-	PN	0.28
store on site 100m distance	m3	-	-	PN	0.08
store on site 200m distance	m3	-	-	PN	0.12

	Unit	Labour grade	Labour hours	Plant grade	Plant hours

Double handling

Load excavated material, transport on site and deposit in new location, distance between stockpiles

	Unit	Labour grade	Labour hours	Plant grade	Plant hours
50m	m3	-	-	PP	0.04
100m	m3	-	-	PP	0.06
200m	m3	-	-	PP	0.08
300m	m3	-	-	PP	0.10
400m	m3	-	-	PP	0.12
500m	m3	-	-	PP	0.14

Load excavated rock, transport on site and deposit in new location, distance between stockpiles

	Unit	Labour grade	Labour hours	Plant grade	Plant hours
50m	m3	-	-	PP	0.06
100m	m3	-	-	PP	0.08
200m	m3	-	-	PP	0.10
300m	m3	-	-	PP	0.12
400m	m3	-	-	PP	0.14
500m	m3	-	-	PP	0.16

Soft spots

Excavate soft spot and replace with

	Unit	Labour grade	Labour hours	Plant grade	Plant hours
granular fill	m3	-	-	PK	0.40
ready mixed concrete	m3	-	-	PK	0.40

	Unit	Labour grade	Labour hours	Plant grade	Plant hours

Filling

Filling to structures

	Unit	Labour grade	Labour hours	Plant grade	Plant hours
Excavated vegetable soil from spoil heap 100m distance	m3	-	-	PO	0.08
Excavated vegetable soil from spoil heap 200m distance	m3	-	-	PO	0.08
Imported vegetable soil	m3	-	-	PO	0.08
Selected subsoil	m3	-	-	PO	0.08
Imported subsoil	m3	-	-	PO	0.08

Filling to embankments

	Unit	Labour grade	Labour hours	Plant grade	Plant hours
Excavated vegetable soil from spoil heap 100m distance	m3	-	-	PO	0.06
Excavated vegetable soil from spoil heap 200m distance	m3	-	-	PO	0.06
Imported vegetable soil	m3	-	-	PO	0.06
Selected subsoil	m3	-	-	PO	0.06
Imported subsoil	m3	-	-	PO	0.06
DTp type 1	m3	-	-	PO	0.06
DTp type 2	m3	-	-	PO	0.06

	Unit	Labour grade	Labour hours	Plant grade	Plant hours
Filling generally					
excavated vegetable soil from spoil heap 100m distance	m3	-	-	PO	0.06
excavated vegetable soil from spoil heap 200m distance	m3	-	-	PO	0.06
imported vegetable soil	m3	-	-	PO	0.06
selected subsoil	m3	-	-	PO	0.06
imported subsoil	m3	-	-	PO	0.06
DTp type 1	m3	-	-	PO	0.06
DTp type 2	m3	-	-	PO	0.06
Filling in layers 100mm thick					
Excavated vegetable soil from spoil heap 100m distance	m2	-	-	PO	0.01
Excavated vegetable soil from spoil heap 200m distance	m2	-	-	PO	0.01
Imported vegetable soil	m2	-	-	PO	0.01
Selected subsoil	m2	-	-	PO	0.01
Imported subsoil	m2	-	-	PO	0.01
DTp type 1	m2	-	-	PO	0.01
DTp type 2	m2	-	-	PO	0.01

	Unit	Labour grade	Labour hours	Plant grade	Plant hours

Filling in layers 200mm thick

	Unit	Labour grade	Labour hours	Plant grade	Plant hours
Excavated vegetable soil from spoil heap 100m distance	m2	-	-	PO	0.01
Excavated vegetable soil from spoil heap 200m distance	m2	-	-	PO	0.01
Imported vegetable soil	m2	-	-	PO	0.01
Selected subsoil	m2	-	-	PO	0.01
Imported subsoil	m2	-	-	PO	0.01
DTp type 1	m2	-	-	PO	0.01
DTp type 2	m2	-	-	PO	0.01

Filling in layers 300mm thick

	Unit	Labour grade	Labour hours	Plant grade	Plant hours
Excavated vegetable soil from spoil heap 100m distance	m2	-	-	PO	0.02
Excavated vegetable soil from spoil heap 200m distance	m2	-	-	PO	0.02
Imported vegetable soil	m2	-	-	PO	0.02
Selected subsoil	m2	-	-	PO	0.02
Imported subsoil	m2	-	-	PO	0.02
DTp type 1	m2	-	-	PO	0.02
DTp type 2	m2	-	-	PO	0.02

	Unit	Labour grade	Labour hours	Plant grade	Plant hours
Filling in layers 400mm thick					
Excavated vegetable soil from spoil heap 100m distance	m2	-	-	PO	0.02
Excavated vegetable soil from spoil heap 200m distance	m2	-	-	PO	0.02
Imported vegetable soil	m2	-	-	PO	0.02
Selected subsoil	m2	-	-	PO	0.02
Imported subsoil	m2	-	-	PO	0.02
DTp type 1	m2	-	-	PO	0.02
DTp type 2	m2	-	-	PO	0.02
Filling in layers 500mm thick					
Excavated vegetable soil from spoil heap 100m distance	m2	-	-	PO	0.03
Excavated vegetable soil from spoil heap 200m distance	m2	-	-	PO	0.03
Imported vegetable soil	m2	-	-	PO	0.03
Selected subsoil	m2	-	-	PO	0.03
Imported subsoil	m2	-	-	PO	0.03
DTp type 1	m2	-	-	PO	0.03
DTp type 2	m2	-	-	PO	0.03

	Unit	Labour grade	Labour hours	Plant grade	Plant hours
Filling sundries					
Trimming excavated surfaces					
horizontally	m2	-	-	PM	0.01
10-45º to the horizontal	m2	-	-	PM	0.01
45-90º to the horizontal	m2	-	-	PM	0.02
Trimming subsoil					
horizontally	m2	-	-	PM	0.01
10-45º to the horizontal	m2	-	-	PM	0.01
45-90º to the horizontal	m2	-	-	PM	0.02
Trimming rock					
horizontally	m2	-	-	PM	0.35
10-45º to the horizontal	m2	-	-	PM	0.40
45-90º to the horizontal	m2	-	-	PM	0.45
Preparation of excavated surfaces					
horizontally	m2	-	-	PM	0.02
10-45º to the horizontal	m2	-	-	PM	0.03
45-90º to the horizontal	m2	-	-	PM	0.04
Preparation of excavated surfaces					
horizontally	m2	-	-	PM	0.02
10-45º to the horizontal	m2	-	-	PM	0.03
45-90º to the horizontal	m2	-	-	PM	0.04
Preparation of rock					
horizontally	m2	-	-	PM	0.40
10-45º to the horizontal	m2	-	-	PM	0.45
45-90º to the horizontal	m2	-	-	PM	0.50

22

Geotextiles

	Roll size m	Roll area m2
Paraweb	50.0 x 3.6	180
Polypropylene	100.0 x 4.5	450
	100.0 x 5.0	500
Stabilising matting	10.0 x 3.0	30
	40.0 x 3.0	120
Erosion control matting	84.0 x 1.2	100.8
Mulch matting	60.0 x 1.2	72

Labour grades

Semi-skilled operative	LB

	Unit	Labour grade	Labour hours	Plant grade	Plant hours
Sheeting					
Paraweb flexible sheeting	m2	LB	0.10	-	-
Polypropylene sheeting					
0.60mm thick	m2	LB	0.08	-	-
0.95mm thick	m2	LB	0.09	-	-
1.20mm thick	m2	LB	0.10	-	-
1.40mm thick	m2	LB	0.11	-	-
1.50mm thick	m2	LB	0.12	-	-
2.50mm thick	m2	LB	0.13	-	-
Matting					
Ground stabilising matting fixed with steel pins					
polypropylene	m2	LB	0.10	-	-
polyethylene	m2	LB	0.12	-	-
Biogradable erosion control mats fixed with steel pins	m2	LB	0.12	-	-

23

Concrete work

Weights of materials	kg/m3
Cement	1440
Sand	1600
Aggregate, coarse	1500
Stone, crushed	1350
Ballast, all-in	1800
Concrete	2450

Suitability of mixes

Precast work in small sectional areas	1:1:2
Watertight reinforced concrete structures	1:1.5:3
Normal reinforced concrete work	1:2:4
Mass unreinforced concrete work	1:2.5:5
Rough concrete work	1:3:6

Concrete mixes (per m3)

Mix	Cement t	Sand m3	Aggregate m3	Water litres
1:1:2	0.50	0.45	0.70	208
1:1.5:3	0.37	0.50	0.80	185
1:2:4	0.30	0.54	0.85	175
1:2.5:5	0.25	0.55	0.85	166
1:3:6	0.22	0.55	0.85	160

Grade

20/20	0.32	0.62	1.20	170
25/20	0.35	0.60	1.17	180
30/20	0.80	0.59	1.11	200
7/40 all-in	0.18	-	1.95	150
20/20 all-in	0.32	-	1.85	170
25/20 all-in	0.36	-	1.75	180

Steel bar reinforcement

Diameter mm	Nominal weight kg/m	Length m/tonne	Sectional area mm2
6	0.222	4505	28.30
8	0.395	2532	50.30
10	0.616	1623	78.50
12	0.888	1126	113.10
16	1.579	633	201.10
20	2.466	406	314.20

Diameter mm	Nominal weight kg/m	Length m/tonne	Sectional area mm2
25	3.854	259	490.90
32	6.313	158	804.20
40	9.864	101	1256.60
50	15.413	65	1963.50

Steel fabric reinforcement

Ref.	Nominal weight kg/m2	Mesh dimensions Main mm	Cross mm	Wire diameters Main mm	Cross mm
A393	6.16	200	200	10	10
A252	3.95	200	200	8	8
A193	3.02	200	200	7	7
A142	2.22	200	200	6	6
A98	1.54	200	200	5	5
B1131	10.90	100	200	12	8
B785	8.14	100	200	10	8
B503	5.93	100	200	8	8
B385	4.53	100	200	7	7
B283	3.73	100	200	6	7
B196	3.05	100	200	5	7
C785	6.72	100	400	10	6
C636	5.55	110	400	9	6
C503	4.34	100	400	8	5
C385	3.41	100	400	7	5
C283	2.61	100	400	6	5
D98	1.54	200	200	5	5
D49	0.77	100	100	2.5	2.5

Formwork stripping times

	Ordinary concrete c.60^0	c.35^0	Rapid hardening concrete c.60^0	c.35^0
Beams, columns, walls	1	1	6	5
Soffits of slabs	3	10	2	7
Soffits of beams	7	12	4	10

Labour grades

1 Ganger, 1 semi-skilled operative
and 1 unskilled operative   LH

2 Craftsmen and 1 unskilled
operative   LI

1 Craftsman and 2 unskilled
operatives   LJ

1 Craftsman and 1 unskilled
operative   LK

Plant grades

1 Crawler crane, 2 concrete skips,
3 vibrating pokers and 1 compressor
(375cfm)   PQ

1 Saw bench and 20% crawler
crane   PR

1 Crawler crane   PS

1 Compressor (375cfm) and 1 tar
boiler   PT

	Unit	Cement tonnes	Sand m3	Aggregate m3	All-in aggregate m3
Provision of site-mixed concrete, nominal mix by volume					
1:1:2	m3	0.50	0.45	0.70	-
1:1.5:3	m3	0.37	0.50	0.80	-
1:2:4	m3	0.30	0.54	0.85	-
1:2.5:3	m3	0.25	0.55	0.85	-
1:3:6	m3	0.22	0.55	0.85	-
1:6	m3	0.31	-	-	0.45
1:9	m3	0.21	-	-	0.52
1:12	m3	0.17	-	-	0.55
Provision of site-mixed concrete, nominal mix by weight					
1:1.5:3	m3	0.22	0.81	1.31	-
1:2:4	m3	0.31	0.81	1.24	-
1:3:6	m3	0.39	0.74	1.17	-
1:6	m3	0.31	-	-	2.15
1:9	m3	0.21	-	-	2.26
1:12	m3	0.17	-	-	2.38

	Unit	Labour grade	Labour hours	Plant grade	Plant hours

Placing unreinforced concrete

Blinding, thickness

	Unit	Labour grade	Labour hours	Plant grade	Plant hours
not exceeding 150mm	m3	LH	0.22	PQ	0.22
150-300mm	m3	LH	0.20	PQ	0.20
300-500mm	m3	LH	0.18	PQ	0.18

Blinding placed against excavated surfaces, thickness

	Unit	Labour grade	Labour hours	Plant grade	Plant hours
not exceeding 150mm	m3	LH	0.25	PQ	0.25
150-300mm	m3	LH	0.23	PQ	0.23
300-500mm	m3	LH	0.11	PQ	0.11

Bases, footings, pile caps and ground slabs, thickness

	Unit	Labour grade	Labour hours	Plant grade	Plant hours
not exceeding 150mm	m3	LH	0.24	PQ	0.24
150-300mm	m3	LH	0.22	PQ	0.22
300-500mm	m3	LH	0.20	PQ	0.20
exceeding 500mm	m3	LH	0.18	PQ	0.18

Walls, thickness

	Unit	Labour grade	Labour hours	Plant grade	Plant hours
not exceeding 150mm	m3	LH	0.18	PQ	0.18
150-300mm	m3	LH	0.16	PQ	0.16
300-500mm	m3	LH	0.14	PQ	0.14
exceeding 500mm	m3	LH	0.12	PQ	0.12

Surrounds to precast concrete chambers, thickness 300mm

	Unit	Labour grade	Labour hours	Plant grade	Plant hours
thickness 300mm	m3	LH	0.22	PQ	0.22

	Unit	Labour grade	Labour hours	Plant grade	Plant hours
Plinths, blocks and the like, size					
750 x 750 x 1000mm	m3	LH	0.20	PQ	0.20
1000 x 1000 x 1000mm	m3	LH	0.18	PQ	0.18
1200 x 1200 x 1000mm	m3	LH	0.16	PQ	0.16

Placing reinforced concrete

Bases, footings, pile caps and ground slabs, thickness					
not exceeding 150mm	m3	LH	0.26	PQ	0.26
150-300mm	m3	LH	0.24	PQ	0.24
300-500mm	m3	LH	0.22	PQ	0.22
exceeding 500mm	m3	LH	0.20	PQ	0.20
Walls, thickness					
not exceeding 150mm	m3	LH	0.24	PQ	0.24
150-300mm	m3	LH	0.22	PQ	0.22
300-500mm	m3	LH	0.20	PQ	0.20
exceeding 500mm	m3	LH	0.18	PQ	0.18
Suspended slabs, thickness					
not exceeding 150mm	m3	LH	0.24	PQ	0.24
150-300mm	m3	LH	0.22	PQ	0.22
300-500mm	m3	LH	0.20	PQ	0.20
exceeding 500mm	m3	LH	0.18	PQ	0.18
Columns and piers, size					
cross-sectional area not exceeding 0.03m2	m3	LH	0.60	PQ	0.60
cross-sectional area 0.03-0.1m2	m3	LH	0.55	PQ	0.55
cross-sectional area 0.1-0.25m2	m3	LH	0.50	PQ	0.50

	Unit	Labour grade	Labour hours	Plant grade	Plant hours
cross-sectional area 0.25-1m2	m3	LH	0.35	PQ	0.35
cross-sectional area exceeding 1m2	m3	LH	0.30	PQ	0.30

Beams

	Unit	Labour grade	Labour hours	Plant grade	Plant hours
cross-sectional area not exceeding 0.03m2	m3	LH	0.60	PQ	0.60
cross-sectional area 0.03-0.1m2	m3	LH	0.55	PQ	0.55
cross-sectional area 0.1-0.25m2	m3	LH	0.50	PQ	0.50
cross-sectional area 0.25-1m2	m3	LH	0.35	PQ	0.35
cross-sectional area exceeding 1m2	m3	LH	0.30	PQ	0.30

Casings to metal sections

	Unit	Labour grade	Labour hours	Plant grade	Plant hours
cross-sectional area not exceeding 0.03m2	m3	LH	0.62	PQ	0.62
cross-sectional area 0.03-0.1m2	m3	LH	0.58	PQ	0.58
cross-sectional area 0.1-0.25m2	m3	LH	0.54	PQ	0.54
cross-sectional area 0.25-1m2	m3	LH	0.38	PQ	0.38
cross-sectional area exceeding 1m2	m3	LH	0.34	PQ	0.34

Plinths, blocks and the like, size

	Unit	Labour grade	Labour hours	Plant grade	Plant hours
750 x 750 x 1000mm	m3	LH	0.22	PQ	0.22
1000 x 1000 x 1000mm	m3	LH	0.20	PQ	0.20
1200 x 1200 x 1000mm	m3	LH	0.18	PQ	0.18

	Unit	Labour grade	Labour hours	Plant grade	Plant hours

Placing prestressed concrete

Suspended slabs, thickness

not exceeding 150mm	m3	LH	0.32	PQ	0.32
150-300mm	m3	LH	0.28	PQ	0.28
300-500mm	m3	LH	0.24	PQ	0.24
exceeding 500mm	m3	LH	0.20	PQ	0.20

Beams

cross-sectional area not exceeding 0.03m2	m3	LH	0.60	PQ	0.60
cross-sectional area 0.03-0.1m2	m3	LH	0.55	PQ	0.55
cross-sectional area 0.1-0.25m2	m3	LH	0.50	PQ	0.50
cross-sectional area 0.25-1m2	m3	LH	0.35	PQ	0.35
cross-sectional area exceeding 1m2	m3	LH	0.30	PQ	0.30

Formwork, rough and fair finish

Plane horizontal, width

not exceeding 0.1m	m	LI	0.15	PR	0.15
0.1-0.2m	m	LI	0.24	PR	0.24
0.2-0.4m	m2	LI	0.48	PR	0.45
0.4-1.22m	m2	LI	0.48	PR	0.45
exceeding 1.22m	m2	LI	0.48	PR	0.45

Plane sloping, width

not exceeding 0.1m	m	LI	0.16	PR	0.16
0.1-0.2m	m	LI	0.25	PR	0.25
0.2-0.4m	m2	LI	0.50	PR	0.50
0.4-1.22m	m2	LI	0.50	PR	0.50
exceeding 1.22m	m2	LI	0.50	PR	0.50

	Unit	Labour grade	Labour hours	Plant grade	Plant hours
Plane battered, width					
not exceeding 0.1m	m	LI	0.18	PR	0.18
0.1-0.2m	m	LI	0.28	PR	0.28
0.2-0.4m	m2	LI	0.54	PR	0.54
0.4-1.22m	m2	LI	0.54	PR	0.54
exceeding 1.22m	m2	LI	0.54	PR	0.54
Plane vertical, width					
not exceeding 0.1m	m	LI	0.18	PR	0.18
0.1-0.2m	m	LI	0.28	PR	0.28
0.2-0.4m	m2	LI	0.54	PR	0.54
0.4-1.22m	m2	LI	0.54	PR	0.54
exceeding 1.22m	m2	LI	0.54	PR	0.54
Curved to one radius in one plane 0.5m radius, width					
not exceeding 0.1m	m	LI	0.22	PR	0.22
0.1-0.2m	m	LI	0.38	PR	0.38
0.2-0.4m	m2	LI	0.75	PR	0.75
0.4-1.22m	m2	LI	0.75	PR	0.75
exceeding 1.22m	m2	LI	0.75	PR	0.75
Curved to one radius in one plane 1m radius, width					
not exceeding 0.1m	m	LI	0.20	PR	0.20
0.1-0.2m	m	LI	0.35	PR	0.35
0.2-0.4m	m2	LI	0.70	PR	0.70
0.4-1.22m	m2	LI	0.70	PR	0.70
exceeding 1.22m	m2	LI	0.70	PR	0.70
Curved to one radius in one plane 1.5m radius, width					
not exceeding 0.1m	m	LI	0.19	PR	0.19
0.1-0.2m	m	LI	0.33	PR	0.33
0.2-0.4m	m2	LI	0.65	PR	0.65
0.4-1.22m	m2	LI	0.65	PR	0.65
exceeding 1.22m	m2	LI	0.65	PR	0.65

	Unit	Labour grade	Labour hours	Plant grade	Plant hours
Curved to one radius in one plane 2m radius, width					
not exceeding 0.1m	m	LI	0.18	PR	0.18
0.1-0.2m	m	LI	0.30	PR	0.30
0.2-0.4m	m2	LI	0.60	PR	0.60
0.4-1.22m	m2	LI	0.60	PR	0.60
exceeding 1.22m	m2	LI	0.60	PR	0.60
To three sides of isolated beams					
100 x 200mm	m	LI	0.40	PR	0.40
100 x 250mm	m	LI	0.40	PR	0.40
100 x 300mm	m	LI	0.40	PR	0.40
150 x 200mm	m	LI	0.40	PR	0.30
200 x 200mm	m	LI	0.40	PR	0.40
200 x 300mm	m	LI	0.50	PR	0.50
300 x 300mm	m	LI	0.50	PR	0.50
300 x 400mm	m	LI	0.52	PR	0.52
300 x 500mm	m	LI	0.54	PR	0.54
400 x 400mm	m	LI	0.56	PR	0.56
400 x 500mm	m	LI	0.56	PR	0.56
500 x 500mm	m	LI	0.60	PR	0.60
To two sides of attached beams					
100 x 200mm	m	LI	0.30	PR	0.30
100 x 250mm	m	LI	0.30	PR	0.30
100 x 300mm	m	LI	0.30	PR	0.30
150 x 200mm	m	LI	0.30	PR	0.30
200 x 200mm	m	LI	0.35	PR	0.35
200 x 300mm	m	LI	0.35	PR	0.35
300 x 300mm	m	LI	0.38	PR	0.38
300 x 400mm	m	LI	0.38	PR	0.38
300 x 500mm	m	LI	0.40	PR	0.40
400 x 400mm	m	LI	0.42	PR	0.42
400 x 500mm	m	LI	0.42	PR	0.42
500 x 500mm	m	LI	0.42	PR	0.42

	Unit	Labour grade	Labour hours	Plant grade	Plant hours
To four sides of isolated columns					
100 x 200mm	m	LI	0.45	PR	0.45
100 x 250mm	m	LI	0.45	PR	0.45
100 x 300mm	m	LI	0.45	PR	0.45
150 x 200mm	m	LI	0.45	PR	0.45
200 x 200mm	m	LI	0.55	PR	0.55
200 x 300mm	m	LI	0.55	PR	0.55
300 x 300mm	m	LI	0.57	PR	0.57
300 x 400mm	m	LI	0.57	PR	0.57
300 x 500mm	m	LI	0.59	PR	0.59
400 x 400mm	m	LI	0.61	PR	0.61
400 x 500mm	m	LI	0.61	PR	0.61
500 x 500mm	m	LI	0.65	PR	0.65
To three sides of attached columns					
100 x 200mm	m	LI	0.40	PR	0.40
100 x 250mm	m	LI	0.40	PR	0.40
100 x 300mm	m	LI	0.40	PR	0.40
150 x 200mm	m	LI	0.40	PR	0.40
200 x 200mm	m	LI	0.45	PR	0.45
200 x 300mm	m	LI	0.45	PR	0.45
300 x 300mm	m	LI	0.47	PR	0.47
300 x 400mm	m	LI	0.47	PR	0.47
300 x 500mm	m	LI	0.59	PR	0.59
400 x 400mm	m	LI	0.61	PR	0.61
400 x 500mm	m	LI	0.61	PR	0.61
500 x 500mm	m	LI	0.65	PR	0.65
To small voids					
not exceeding 0.5m	nr	LI	0.20	-	-
0.5-1m	nr	LI	0.25	-	-
1-2m	nr	LI	0.30	-	-
2-5m	nr	LI	0.35	-	-

	Unit	Labour grade	Labour hours	Plant grade	Plant hours
To large voids					
not exceeding 0.5m	nr	LI	0.30	-	-
0.5-1m	nr	LI	0.35	-	-
1-2m	nr	LI	0.40	-	-
2-5m	nr	LI	0.45	-	-

Bar reinforcement

Mild steel round bars in straight lengths, nominal size

	Unit	Labour grade	Labour hours	Plant grade	Plant hours
6mm	t	LJ	9.50	PS	9.50
8mm	t	LJ	9.00	PS	9.00
10mm	t	LJ	8.50	PS	8.50
12mm	t	LJ	7.50	PS	7.50
16mm	t	LJ	7.00	PS	7.00
20mm	t	LJ	6.00	PS	6.00
25mm	t	LJ	5.00	PS	5.00
32mm	t	LJ	4.50	PS	4.50
40mm	t	LJ	4.00	PS	4.00

Mild steel round bars in bent lengths, nominal size

	Unit	Labour grade	Labour hours	Plant grade	Plant hours
6mm	t	LJ	9.50	PS	9.50
8mm	t	LJ	9.00	PS	9.00
10mm	t	LJ	8.50	PS	8.50
12mm	t	LJ	7.50	PS	7.50
16mm	t	LJ	7.00	PS	7.00
20mm	t	LJ	6.00	PS	6.00
25mm	t	LJ	5.00	PS	5.00
32mm	t	LJ	4.50	PS	4.50
40mm	t	LJ	4.00	PS	4.00

	Unit	Labour grade	Labour hours	Plant grade	Plant hours

Deformed high yield steel
bars in straight lengths,
nominal size

	Unit	Labour grade	Labour hours	Plant grade	Plant hours
6mm	t	LJ	9.50	PS	9.50
8mm	t	LJ	9.00	PS	9.00
10mm	t	LJ	8.50	PS	8.50
12mm	t	LJ	7.50	PS	7.50
16mm	t	LJ	7.00	PS	7.00
20mm	t	LJ	6.00	PS	6.00
25mm	t	LJ	5.00	PS	5.00
32mm	t	LJ	4.50	PS	4.50
40mm	t	LJ	4.00	PS	4.00

Deformed high yield steel
bars in bent lengths,nominal size

	Unit	Labour grade	Labour hours	Plant grade	Plant hours
6mm	t	LJ	9.50	PS	9.50
8mm	t	LJ	9.00	PS	9.00
10mm	t	LJ	8.50	PS	8.50
12mm	t	LJ	7.50	PS	7.50
16mm	t	LJ	7.00	PS	7.00
20mm	t	LJ	5.00	PS	6.00
25mm	t	LJ	5.00	PS	5.00
32mm	t	LJ	4.50	PS	4.50
40mm	t	LJ	4.00	PS	4.00

Mesh reinforcement

Steel fabric mesh reinforcement,
nominal mass not exceeding
2kg/m2

	Unit	Labour grade	Labour hours	Plant grade	Plant hours
D49	m2	LJ	0.05	PS	0.05
A98	m2	LJ	0.05	PS	0.05

Steel fabric mesh reinforcement,
nominal mass not exceeding
2-3kg/m2

	Unit	Labour grade	Labour hours	Plant grade	Plant hours
A142	m2	LJ	0.06	PS	0.06
C283	m2	LJ	0.06	PS	0.06

	Unit	Labour grade	Labour hours	Plant grade	Plant hours
Steel fabric mesh reinforcement, nominal mass not exceeding 3-4kg/m2					
A193	m2	LJ	0.07	PS	0.07
B196	m2	LJ	0.07	PS	0.07
C385	m2	LJ	0.07	PS	0.07
B283	m2	LJ	0.07	PS	0.07
A252	m2	LJ	0.07	PS	0.07
Steel fabric mesh reinforcement, nominal mass not exceeding 4-5kg/m2					
C503	m2	LJ	0.08	PS	0.08
B385	m2	LJ	0.08	PS	0.08
Steel fabric mesh reinforcement, nominal mass not exceeding 5-6kg/m2					
C636	m2	LJ	0.10	PS	0.10
B503	m2	LJ	0.10	PS	0.10
Steel fabric mesh reinforcement, nominal mass not exceeding 6-7kg/m2					
A393	m2	LJ	0.14	PS	0.14
C785	m2	LJ	0.14	PS	0.14

Joints

Open plain, average width

	Unit	Labour grade	Labour hours	Plant grade	Plant hours
not exceeding 0.5m	m2	LK	0.08	PT	0.08
0.5-1m	m2	LK	0.07	PT	0.07
1-1.5m	m2	LK	0.06	PT	0.06

	Unit	Labour grade	Labour hours	Plant grade	Plant hours
Open with joint filler, 12mm thick, average width					
not exceeding 0.5m	m2	LK	0.18	PT	0.18
0.5-1m	m2	LK	0.15	PT	0.15
1-1.5m	m2	LK	0.13	PT	0.13
Open with joint filler, 19mm thick, average width					
not exceeding 0.5m	m2	LK	0.22	PT	0.22
0.5-1m	m2	LK	0.18	PT	0.18
1-1.5m	m2	LK	0.15	PT	0.15
Open with joint filler, 25mm thick, average width					
not exceeding 0.5m	m2	LK	0.25	PT	0.25
0.5-1m	m2	LK	0.20	PT	0.20
1-1.5m	m2	LK	0.15	PT	0.15
Formed surface plain joint including formwork, average width					
not exceeding 0.5m	m2	LK	0.45	PT	0.45
0.5-1m	m2	LK	0.40	PT	0.40
1-1.5m	m2	LK	0.35	PT	0.35
Formed surface joint including formwork with filler 12mm thick, average width					
not exceeding 0.5m	m2	LK	0.55	PT	0.55
0.5-1m	m2	LK	0.50	PT	0.50
1-1.5m	m2	LK	0.45	PT	0.45

	Unit	Labour grade	Labour hours	Plant grade	Plant hours
Formed surface joint including formwork with filler 19mm thick, average width					
not exceeding 0.5m	m2	LK	0.60	PT	0.60
0.5-1m	m2	LK	0.55	PT	0.55
1-1.5m	m2	LK	0.50	PT	0.50
Formed surface joint including formwork with filler 25mm thick, average width					
not exceeding 0.5m	m2	LK	0.65	PT	0.65
0.5-1m	m2	LK	0.60	PT	0.60
1-1.5m	m2	LK	0.55	PT	0.55

Waterstops

PVC-U flat dumbbell waterstop, width

	Unit	Labour grade	Labour hours	Plant grade	Plant hours
not exceeding 150mm	m	LK	0.14	-	-
150-200mm	m	LK	0.15	-	-
200-300mm	m	LK	0.16	-	-
junction piece	m	LK	0.25	-	-

PVC-U centre bulb waterstop, width

	Unit	Labour grade	Labour hours	Plant grade	Plant hours
not exceeding 150mm	m	LK	0.14	-	-
150-200mm	m	LK	0.15	-	-
200-300mm	m	LK	0.16	-	-
junction piece	nr	LK	0.25	-	-

Rubber flat dumbbell waterstop, width

	Unit	Labour grade	Labour hours	Plant grade	Plant hours
not exceeding 150mm	m	LK	0.14	-	-
150-200mm	m	LK	0.15	-	-
200-300mm	m	LK	0.16	-	-
junction piece	nr	LK	0.25	-	-

	Unit	Labour grade	Labour hours	Plant grade	Plant hours
Rubber centre bulb waterstop, width					
not exceeding 150mm	m	LK	0.14	-	-
150-200mm	m	LK	0.15	-	-
200-300mm	m	LK	0.16	-	-
junction piece	nr	LK	0.25	-	-

Rebates

Sealed rebate or groove with hot poured bitumen sealing compound, size

	Unit	Labour grade	Labour hours	Plant grade	Plant hours
10 x 10mm	m	LK	0.06	-	-
10 x 20mm	m	LK	0.06	-	-
10 x 25mm	m	LK	0.06	-	-
15 x 15mm	m	LK	0.07	-	-
15 x 20mm	m	LK	0.07	-	-
15 x 25mm	m	LK	0.08	-	-
20 x 20mm	m	LK	0.08	-	-
20 x 30mm	m	LK	0.09	-	-
20 x 40mm	m	LK	0.09	-	-

Dowels

Plain mild steel, length 500mm, cast into side of joint, diameter

	Unit	Labour grade	Labour hours	Plant grade	Plant hours
12mm	m	LK	0.24	-	-
16mm	m	LK	0.26	-	-
20mm	m	LK	0.28	-	-
25mm	m	LK	0.30	-	-

Plain mild steel, length 750mm, cast into side of joint, diameter

	Unit	Labour grade	Labour hours	Plant grade	Plant hours
12mm	m	LK	0.26	-	-
16mm	m	LK	0.30	-	-
20mm	m	LK	0.32	-	-
25mm	m	LK	0.34	-	-

	Unit	Labour grade	Labour hours	Plant grade	Plant hours
Plain mild steel, length 750mm, cast into side of joint, diameter					
12mm	m	LK	0.30	-	-
16mm	m	LK	0.34	-	-
20mm	m	LK	0.36	-	-
25mm	m	LK	0.38	-	-

Concrete finishes

	Unit	Labour grade	Labour hours	Plant grade	Plant hours
Wood float finish to concrete					
level	m2	LK	0.06	-	-
falls and crossfalls	m2	LK	0.08	-	-
Steel trowel finish to concrete					
level	m2	LK	0.06	-	-
falls and crossfalls	m2	LK	0.08	-	-
Granolithic finish 19mm thick to concrete					
level	m2	LK	0.07	-	-
falls and crossfalls	m2	LK	0.09	-	-
Granolithic finish 25mm thick to concrete					
level	m2	LK	0.07	-	-
falls and crossfalls	m2	LK	0.09	-	-
Aggregate exposed to concrete retarder	m2	LK	0.18	-	-
Bush hammering face of concrete	m2	LK	0.40	-	-

	Unit	Labour grade	Labour hours	Plant grade	Plant hours
Rubbing down face of concrete after striking concrete	m2	LK	0.18	-	-

Inserts

	Unit	Labour grade	Labour hours	Plant grade	Plant hours
PVC-U conduit, 100mm diameter	nr	LK	0.75	-	-
PVC-U conduit, 100mm diameter	nr	LK	0.85	-	-
Mild steel pipe, 150mm diameter x 450mm length projecting from two surfaces	nr	LK	1.00	-	-

Grouting

Grout under base plates
with cement mortar (1:3)
25mm thick, area

	Unit	Labour grade	Labour hours	Plant grade	Plant hours
not exceeding 0.1m2	nr	LK	0.55	-	-
0.1-0.5m2	nr	LK	0.60	-	-
0.5-1m2	nr	LK	0.70	-	-
1-1.5m2	nr	LK	0.75	-	-
1.5-2m2	nr	LK	0.80	-	-

24

Drainage

Weights of materials		kg/m3
Ashes		800
Bricks, common		1760
engineering		1760
Cement		1900
Concrete		2300
Gravel		1750
Limestone, crushed		1760
Sand		1600

	Diameter	**kg/m**
PVC-U pipes	80mm	1.20
	110mm	1.60
	160mm	3.00
	200mm	4.60
	250mm	7.20
Vitrified clay pipes	100mm	15.63
	150mm	37.04
	225mm	95.24
	300mm	196.08
	400mm	357.14
	450mm	500.00
	500mm	555.60
Unreinforced concrete pipes	300mm	83.00
	375mm	115.00
	450mm	144.00
	525mm	197.00
	600mm	240.00
	675mm	283.00
	750mm	355.00
	825mm	402.00

Diameter	kg/m
900mm	473.00
975mm	529.00
1050mm	610.00
1125mm	732.00
1200mm	796.00
1350mm	1100.00
1500mm	1222.00

Spun iron spigot and socket pipes		
	100mm	22.04
	150mm	38.15
	225mm	68.33
	300mm	105.93
	375mm	147.03
	450mm	193.15
	525mm	236.11
	600mm	295.93
	675mm	371.30

Steel pipes		
	50mm	4.20
	75mm	7.84
	100mm	10.05
	150mm	17.83
	225mm	31.05
	300mm	41.75
	450mm	64.06
	600mm	121.11
	750mm	168.00
	900mm	219.00
	1200mm	340.00
	1500mm	544.00

Volumes of filling (m3/m)

Pipe dia. mm	Beds 50mm	100mm	150mm	Bed and haunching	Surround
100	0.023	0.045	0.068	0.117	0.185
150	0.026	0.053	0.079	0.152	0.231
225	0.030	0.060	0.090	0.195	0.285
300	0.038	0.075	0.113	0.279	0.391
400	-	0.105	0.120	0.285	0.438

Pipe dia. mm	Beds 50mm	100mm	150mm	Bed and haunching	Surround
500	-	0.130	0.128	0.315	0.483
600	-	-	0.155	0.346	0.635
675	-	-	0.170	0.380	0.706
750	-	-	0.180	0.427	0.791
825	-	-	0.195	0.488	0.892
900	-	-	0.206	0.528	0.960
975	-	-	0.216	0.595	1.050
1050	-	-	0.228	0.632	1.162
1125	-	-	0.242	0.675	1.240
1200	-	-	0.259	0.725	1.361
1350	-	-	0.285	0.656	1.515
1500	-	-	0.311	0.953	1.798

Trench widths

Pipe dia. mm	Less than 1.5m deep mm	More than 1.5m deep mm
100	450	600
150	500	650
225	600	750
300	650	800
400	750	900
450	900	1050
600	1000	1300
675	-	1375
750	-	1550
825	-	1625
900	-	1700
975	-	1875
1050	-	1950
1125	-	2025
1200	-	2200
1350	-	2350
1500	-	2500

Labour grades

1 Ganger, 2 semi-skilled
operatives and 1 unskilled
operative   LL

1 Ganger, 1 skilled operative,
2 semi-skilled operatives and
1 unskilled operative   LM

Plant grades

Wheeled hydraulic excavator
(1.7m3), 1 pump (170m3/h),
50 trench sheets, 50 props,
1 dumper (1.5t) and 1 vibratory
compactor   PU

Crawler hydraulic excavator
(1.7m3), 1 pump (275m3/h),
125 trench sheets, 100 props,
1 dumper (1.5t) and 1 vibratory
compactor   PV

Wheeled hydraulic excavator
(1.7m3), 1 dumper (1.5t) and
1 pump (170m3/h)   PW

	Unit	Labour grade	Labour hours	Plant grade	Plant hours

Pipework

Vitrified clay spigot
and socket pipes with
cement mortar to joints,
100mm nominal bore

	Unit	Labour grade	Labour hours	Plant grade	Plant hours
not in trenches	m	LL	0.03	PU	0.03
not exceeding 1.5m deep	m	LL	0.10	PU	0.10
1.5-2.0m deep	m	LL	0.16	PU	0.16
2.0-2.5m deep	m	LL	0.20	PU	0.20
2.5-3.0m deep	m	LL	0.25	PU	0.25
3.0-3.5m deep	m	LL	0.32	PU	0.32
3.5-4.0m deep	m	LO	0.45	PV	0.45
4.0-4.5m deep	m	LO	0.55	PV	0.55
4.5-5.0m deep	m	LO	0.65	PV	0.65
5.0-5.5m deep	m	LO	0.85	PV	0.85
5.5-6.0m deep	m	LO	1.00	PV	1.00

Vitrified clay spigot
and socket pipes with
cement mortar to joints,
150mm nominal bore

	Unit	Labour grade	Labour hours	Plant grade	Plant hours
not in trenches	m	LL	0.05	PU	0.05
not exceeding 1.5m deep	m	LL	0.12	PU	0.12
1.5-2.0m deep	m	LL	0.18	PU	0.18
2.0-2.5m deep	m	LL	0.22	PU	0.22
2.5-3.0m deep	m	LL	0.25	PU	0.25
3.0-3.5m deep	m	LL	0.35	PU	0.35
3.5-4.0m deep	m	LO	0.48	PV	0.48
4.0-4.5m deep	m	LO	0.58	PV	0.58
4.5-5.0m deep	m	LO	0.68	PV	0.68
5.0-5.5m deep	m	LO	0.90	PV	0.90
5.5-6.0m deep	m	LO	1.05	PV	1.05

	Unit	Labour grade	Labour hours	Plant grade	Plant hours

Vitrified clay spigot and socket pipes with cement mortar to joints, 225mm nominal bore

	Unit	Labour grade	Labour hours	Plant grade	Plant hours
not in trenches	m	LL	0.08	PU	0.08
not exceeding 1.5m deep	m	LL	0.14	PU	0.14
1.5-2.0m deep	m	LL	0.20	PU	0.20
2.0-2.5m deep	m	LL	0.24	PU	0.24
2.5-3.0m deep	m	LL	0.30	PU	0.30
3.0-3.5m deep	m	LL	0.38	PU	0.38
3.5-4.0m deep	m	LO	0.50	PV	0.50
4.0-4.5m deep	m	LO	0.60	PV	0.60
4.5-5.0m deep	m	LO	0.70	PV	0.70
5.0-5.5m deep	m	LO	0.95	PV	0.95
5.5-6.0m deep	m	LO	1.10	PV	1.10

Vitrified clay spigot and socket pipes with cement mortar to joints, 300mm nominal bore

	Unit	Labour grade	Labour hours	Plant grade	Plant hours
not in trenches	m	LL	0.10	PU	0.10
not exceeding 1.5m deep	m	LL	0.16	PU	0.16
1.5-2.0m deep	m	LL	0.22	PU	0.22
2.0-2.5m deep	m	LL	0.26	PU	0.26
2.5-3.0m deep	m	LL	0.32	PU	0.32
3.0-3.5m deep	m	LL	0.40	PU	0.40
3.5-4.0m deep	m	LO	0.53	PV	0.53
4.0-4.5m deep	m	LO	0.63	PV	0.63
4.5-5.0m deep	m	LO	0.75	PV	0.75
5.0-5.5m deep	m	LO	1.00	PV	1.00
5.5-6.0m deep	m	LO	1.15	PV	1.15

	Unit	Labour grade	Labour hours	Plant grade	Plant hours

Vitrified clay spigot
and socket pipes with
cement mortar to joints,
375mm nominal bore

	Unit	Labour grade	Labour hours	Plant grade	Plant hours
not in trenches	m	LL	0.12	PU	0.12
not exceeding 1.5m deep	m	LL	0.18	PU	0.18
1.5-2.0m deep	m	LL	0.24	PU	0.24
2.0-2.5m deep	m	LL	0.28	PU	0.28
2.5-3.0m deep	m	LL	0.34	PU	0.34
3.0-3.5m deep	m	LL	0.42	PU	0.42
3.5-4.0m deep	m	LO	0.55	PV	0.55
4.0-4.5m deep	m	LO	0.65	PV	0.65
4.5-5.0m deep	m	LO	0.80	PV	0.80
5.0-5.5m deep	m	LO	1.05	PV	1.05
5.5-6.0m deep	m	LO	1.20	PV	1.20

Vitrified clay spigot
and socket pipes with
cement mortar to joints,
400mm nominal bore

	Unit	Labour grade	Labour hours	Plant grade	Plant hours
not in trenches	m	LL	0.14	PU	0.14
not exceeding 1.5m deep	m	LL	0.20	PU	0.20
1.5-2.0m deep	m	LL	0.26	PU	0.26
2.0-2.5m deep	m	LL	0.30	PU	0.30
2.5-3.0m deep	m	LL	0.36	PU	0.36
3.0-3.5m deep	m	LL	0.45	PU	0.45
3.5-4.0m deep	m	LO	0.60	PV	0.60
4.0-4.5m deep	m	LO	0.70	PV	0.70
4.5-5.0m deep	m	LO	0.85	PV	0.85
5.0-5.5m deep	m	LO	1.10	PV	1.10
5.5-6.0m deep	m	LO	1.25	PV	1.25

	Unit	Labour grade	Labour hours	Plant grade	Plant hours
Vitrified clay spigot and socket pipes with cement mortar to joints, 450mm nominal bore					
not in trenches	m	LL	0.16	PU	0.16
not exceeding 1.5m deep	m	LL	0.22	PU	0.20
1.5-2.0m deep	m	LL	0.28	PU	0.28
2.0-2.5m deep	m	LL	0.32	PU	0.32
2.5-3.0m deep	m	LL	0.38	PU	0.38
3.0-3.5m deep	m	LL	0.48	PU	0.48
3.5-4.0m deep	m	LO	0.63	PV	0.64
4.0-4.5m deep	m	LO	0.74	PV	0.74
4.5-5.0m deep	m	LO	0.90	PV	0.90
5.0-5.5m deep	m	LO	1.15	PV	1.15
5.5-6.0m deep	m	LO	1.30	PV	1.30
Concrete vibrated pipes with flexible joints, Class L, 300mm nominal bore					
not in trenches	m	LL	0.10	PU	0.10
not exceeding 1.5m deep	m	LL	0.18	PU	0.18
1.5-2.0m deep	m	LL	0.23	PU	0.23
2.0-2.5m deep	m	LL	0.30	PU	0.30
2.5-3.0m deep	m	LL	0.38	PU	0.38
3.0-3.5m deep	m	LL	0.43	PU	0.43
3.5-4.0m deep	m	LO	0.45	PV	0.45
4.0-4.5m deep	m	LO	0.47	PV	0.47
4.5-5.0m deep	m	LO	0.60	PV	0.60
5.0-5.5m deep	m	LO	0.70	PV	0.70
5.5-6.0m deep	m	LO	0.80	PV	0.80

	Unit	Labour grade	Labour hours	Plant grade	Plant hours
Concrete vibrated pipes with flexible joints, Class L, 375mm nominal bore					
not in trenches	m	LL	0.12	PU	0.12
not exceeding 1.5m deep	m	LL	0.20	PU	0.20
1.5-2.0m deep	m	LL	0.25	PU	0.25
2.0-2.5m deep	m	LL	0.32	PU	0.32
2.5-3.0m deep	m	LL	0.40	PU	0.40
3.0-3.5m deep	m	LL	0.45	PU	0.45
3.5-4.0m deep	m	LO	0.47	PV	0.47
4.0-4.5m deep	m	LO	0.49	PV	0.49
4.5-5.0m deep	m	LO	0.62	PV	0.62
5.0-5.5m deep	m	LO	0.72	PV	0.72
Concrete vibrated pipes with flexible joints, Class L, 450mm nominal bore					
not in trenches	m	LL	0.14	PU	0.14
not exceeding 1.5m deep	m	LL	0.22	PU	0.22
1.5-2.0m deep	m	LL	0.27	PU	0.27
2.0-2.5m deep	m	LL	0.32	PU	0.32
2.5-3.0m deep	m	LL	0.42	PU	0.42
3.0-3.5m deep	m	LL	0.46	PU	0.46
3.5-4.0m deep	m	LO	0.48	PV	0.48
4.0-4.5m deep	m	LO	0.50	PV	0.50
4.5-5.0m deep	m	LO	0.65	PV	0.65
5.0-5.5m deep	m	LO	0.74	PV	0.74
5.5-6.0m deep	m	LO	1.18	PV	1.18

	Unit	Labour grade	Labour hours	Plant grade	Plant hours
Concrete vibrated pipes with flexible joints, Class L, 525mm nominal bore					
not in trenches	m	LL	0.16	PU	0.16
not exceeding 1.5m deep	m	LL	0.25	PU	0.25
1.5-2.0m deep	m	LL	0.28	PU	0.28
2.0-2.5m deep	m	LL	0.35	PU	0.35
2.5-3.0m deep	m	LL	0.44	PU	0.44
3.0-3.5m deep	m	LL	0.48	PU	0.48
3.5-4.0m deep	m	LO	0.50	PV	0.50
4.0-4.5m deep	m	LO	0.52	PV	0.52
4.5-5.0m deep	m	LO	0.67	PV	0.67
5.0-5.5m deep	m	LO	0.80	PV	0.80
5.5-6.0m deep	m	LO	1.20	PV	1.20
Concrete vibrated pipes with flexible joints, Class L, 600mm nominal bore					
not in trenches	m	LL	0.18	PU	0.18
not exceeding 1.5m deep	m	LL	0.27	PU	0.27
1.5-2.0m deep	m	LL	0.30	PU	0.30
2.0-2.5m deep	m	LL	0.37	PU	0.37
2.5-3.0m deep	m	LL	0.46	PU	0.46
3.0-3.5m deep	m	LL	0.55	PU	0.55
3.5-4.0m deep	m	LO	0.60	PV	0.60
4.0-4.5m deep	m	LO	0.67	PV	0.67
4.5-5.0m deep	m	LO	0.75	PV	0.75
5.0-5.5m deep	m	LO	0.84	PV	0.84
5.5-6.0m deep	m	LO	1.25	PV	1.25

	Unit	Labour grade	Labour hours	Plant grade	Plant hours
Concrete spun pipes with flexible joints, Class L, 750mm nominal bore					
not in trenches	m	LL	0.20	PU	0.20
not exceeding 1.5m deep	m	LL	0.30	PU	0.30
1.5-2.0m deep	m	LL	0.35	PU	0.35
2.0-2.5m deep	m	LL	0.40	PU	0.40
2.5-3.0m deep	m	LL	0.45	PU	0.46
3.0-3.5m deep	m	LL	0.55	PU	0.55
3.5-4.0m deep	m	LO	0.65	PV	0.65
4.0-4.5m deep	m	LO	0.70	PV	0.70
4.5-5.0m deep	m	LO	0.85	PV	0.85
5.0-5.5m deep	m	LO	1.05	PV	1.05
5.5-6.0m deep	m	LO	1.20	PV	1.20
Concrete spun pipes with flexible joints, Class L, 900mm nominal bore					
not in trenches	m	LL	0.22	PU	0.22
not exceeding 1.5m deep	m	LL	0.40	PU	0.40
1.5-2.0m deep	m	LL	0.45	PV	0.45
2.0-2.5m deep	m	LL	0.50	PU	0.50
2.5-3.0m deep	m	LL	0.60	PU	0.60
3.0-3.5m deep	m	LL	0.65	PU	0.65
3.5-4.0m deep	m	LO	0.70	PV	0.70
4.0-4.5m deep	m	LO	0.80	PV	0.80
4.5-5.0m deep	m	LO	0.95	PV	0.95
5.0-5.5m deep	m	LO	1.20	PV	1.20
5.5-6.0m deep	m	LO	1.35	PV	1.35

	Unit	Labour grade	Labour hours	Plant grade	Plant hours

Concrete spun pipes with
flexible joints, Class L,
1200mm nominal bore

	Unit	Labour grade	Labour hours	Plant grade	Plant hours
not in trenches	m	LL	0.24	PU	0.24
not exceeding 1.5m deep	m	LL	0.50	PU	0.50
1.5-2.0m deep	m	LL	0.55	PU	0.55
2.0-2.5m deep	m	LL	0.65	PU	0.65
2.5-3.0m deep	m	LL	0.75	PU	0.75
3.0-3.5m deep	m	LL	0.80	PU	0.80
3.5-4.0m deep	m	LO	0.83	PV	0.83
4.0-4.5m deep	m	LO	0.90	PV	0.90
4.5-5.0m deep	m	LO	1.05	PV	1.05
5.0-5.5m deep	m	LO	1.35	PV	1.35
5.5-6.0m deep	m	LO	1.45	PV	1.45

Concrete spun pipes with
flexible joints, Class L,
1500mm nominal bore

	Unit	Labour grade	Labour hours	Plant grade	Plant hours
not in trenches	m	LL	0.28	PU	0.28
not exceeding 1.5m deep	m	LL	0.70	PU	0.50
1.5-2.0m deep	m	LL	0.80	PU	0.55
2.0-2.5m deep	m	LL	0.90	PU	0.65
2.5-3.0m deep	m	LL	0.75	PU	0.75
3.0-3.5m deep	m	LL	0.80	PU	0.80
3.5-4.0m deep	m	LO	0.83	PV	0.83
4.0-4.5m deep	m	LO	0.90	PV	0.90
4.5-5.0m deep	m	LO	1.05	PV	1.05
5.0-5.5m deep	m	LO	1.35	PV	1.35
5.5-6.0m deep	m	LO	1.45	PV	1.45

	Unit	Labour grade	Labour hours	Plant grade	Plant hours
Concrete spun pipes with flexible joints, Class L, 1800mm nominal bore					
not in trenches	m	LL	0.36	PU	0.36
not exceeding 1.5m deep	m	LL	0.75	PU	0.75
1.5-2.0m deep	m	LL	0.90	PU	0.90
2.0-2.5m deep	m	LL	1.05	PU	1.05
2.5-3.0m deep	m	LL	1.15	PU	1.15
3.0-3.5m deep	m	LL	1.20	PU	1.20
3.5-4.0m deep	m	LO	1.00	PV	1.00
4.0-4.5m deep	m	LO	1.20	PV	1.20
4.5-5.0m deep	m	LO	1.65	PV	1.65
5.0-5.5m deep	m	LO	1.75	PV	1.75
5.5-6.0m deep	m	LO	1.95	PV	1.95
Concrete spun pipes with flexible joints, Class L, 2100mm nominal bore					
not in trenches	m	LL	0.46	PU	0.46
not exceeding 1.5m deep	m	LL	1.20	PU	1.20
1.5-2.0m deep	m	LL	1.20	PU	1.20
2.0-2.5m deep	m	LL	1.35	PU	1.35
2.5-3.0m deep	m	LL	1.50	PU	1.50
3.0-3.5m deep	m	LL	1.60	PU	1.60
3.5-4.0m deep	m	LO	1.26	PV	1.26
4.0-4.5m deep	m	LO	1.50	PV	1.50
4.5-5.0m deep	m	LO	1.80	PV	1.80
5.0-5.5m deep	m	LO	2.10	PV	2.10
5.5-6.0m deep	m	LO	2.25	PV	2.25

	Unit	Labour grade	Labour hours	Plant grade	Plant hours
Ductile iron spigot and socket pipes with Stanlock joints, 150mm nominal bore					
not in trenches	m	LL	0.07	PU	0.07
not exceeding 1.5m deep	m	LL	0.20	PU	0.20
1.5-2.0m deep	m	LL	0.28	PU	0.28
2.0-2.5m deep	m	LL	0.35	PU	0.35
2.5-3.0m deep	m	LL	0.40	PU	0.40
3.0-3.5m deep	m	LL	0.42	PU	0.42
3.5-4.0m deep	m	LO	0.48	PV	0.48
4.0-4.5m deep	m	LO	0.50	PV	0.50
4.5-5.0m deep	m	LO	0.65	PV	0.65
5.0-5.5m deep	m	LO	0.75	PV	0.75
5.5-6.0m deep	m	LO	0.85	PV	0.85
Ductile iron spigot and socket pipes with Stanlock joints, 200mm nominal bore					
not in trenches	m	LL	0.10	PU	0.10
not exceeding 1.5m deep	m	LL	0.25	PU	0.25
1.5-2.0m deep	m	LL	0.35	PU	0.35
2.0-2.5m deep	m	LL	0.40	PU	0.40
2.5-3.0m deep	m	LL	0.45	PU	0.45
3.0-3.5m deep	m	LL	0.48	PU	0.48
3.5-4.0m deep	m	LO	0.50	PV	0.50
4.0-4.5m deep	m	LO	0.55	PV	0.55
4.5-5.0m deep	m	LO	0.70	PV	0.70
5.0-5.5m deep	m	LO	0.85	PV	0.85
5.5-6.0m deep	m	LO	0.95	PV	0.95

	Unit	Labour grade	Labour hours	Plant grade	Plant hours

Ductile iron spigot and
socket pipes with
Stanlock joints,
450mm nominal bore

	Unit	Labour grade	Labour hours	Plant grade	Plant hours
not in trenches	m	LL	0.12	PU	0.12
not exceeding 1.5m deep	m	LL	0.35	PU	0.35
1.5-2.0m deep	m	LL	0.44	PU	0.44
2.0-2.5m deep	m	LL	0.50	PU	0.50
2.5-3.0m deep	m	LL	0.65	PU	0.65
3.0-3.5m deep	m	LL	0.70	PU	0.70
3.5-4.0m deep	m	LO	0.75	PV	0.75
4.0-4.5m deep	m	LO	0.80	PV	0.80
4.5-5.0m deep	m	LO	0.90	PV	0.90
5.0-5.5m deep	m	LO	1.10	PV	1.10
5.5-6.0m deep	m	LO	1.15	PV	1.15

Ductile iron spigot and
socket pipes with
Stanlock joints,
600mm nominal bore

	Unit	Labour grade	Labour hours	Plant grade	Plant hours
not in trenches	m	LL	0.16	PU	0.16
not exceeding 1.5m deep	m	LL	0.40	PU	0.40
1.5-2.0m deep	m	LL	0.50	PU	0.50
2.0-2.5m deep	m	LL	0.60	PU	0.60
2.5-3.0m deep	m	LL	0.70	PU	0.70
3.0-3.5m deep	m	LL	0.75	PU	0.75
3.5-4.0m deep	m	LO	0.80	PV	0.80
4.0-4.5m deep	m	LO	0.85	PV	0.85
4.5-5.0m deep	m	LO	1.00	PV	1.00
5.0-5.5m deep	m	LO	1.20	PV	1.20
5.5-6.0m deep	m	LO	1.25	PV	1.25

	Unit	Labour grade	Labour hours	Plant grade	Plant hours
Steel pipes with electric resistant welded joints 100mm nominal bore					
not in trenches	m	LL	0.06	PU	0.06
not exceeding 1.5m deep	m	LL	0.16	PU	0.16
1.5-2.0m deep	m	LL	0.20	PU	0.20
2.0-2.5m deep	m	LL	0.24	PU	0.24
2.5-3.0m deep	m	LL	0.28	PU	0.28
3.0-3.5m deep	m	LL	0.32	PU	0.32
3.5-4.0m deep	m	LO	0.36	PV	0.36
4.0-4.5m deep	m	LO	0.40	PV	0.40
4.5-5.0m deep	m	LO	0.44	PV	0.44
5.0-5.5m deep	m	LO	0.48	PV	0.48
5.5-6.0m deep	m	LO	0.52	PV	0.52
Steel pipes with electric resistant welded joints 125mm nominal bore					
not in trenches	m	LL	0.08	PU	0.08
not exceeding 1.5m deep	m	LL	0.20	PU	0.20
1.5-2.0m deep	m	LL	0.24	PU	0.24
2.0-2.5m deep	m	LL	0.28	PU	0.28
2.5-3.0m deep	m	LL	0.32	PU	0.32
3.0-3.5m deep	m	LL	0.36	PU	0.36
3.5-4.0m deep	m	LO	0.40	PV	0.40
4.0-4.5m deep	m	LO	0.44	PV	0.44
4.5-5.0m deep	m	LO	0.48	PV	0.48
5.0-5.5m deep	m	LO	0.52	PV	0.52
5.5-6.0m deep	m	LO	0.56	PV	0.56

	Unit	Labour grade	Labour hours	Plant grade	Plant hours
Steel pipes with electric resistant welded joints 150mm nominal bore					
not in trenches	m	LL	0.12	PU	0.12
not exceeding 1.5m deep	m	LL	0.24	PU	0.24
1.5-2.0m deep	m	LL	0.28	PU	0.28
2.0-2.5m deep	m	LL	0.32	PU	0.32
2.5-3.0m deep	m	LL	0.36	PU	0.36
3.0-3.5m deep	m	LL	0.40	PU	0.40
3.5-4.0m deep	m	LO	0.44	PV	0.44
4.0-4.5m deep	m	LO	0.48	PV	0.48
4.5-5.0m deep	m	LO	0.52	PV	0.52
5.0-5.5m deep	m	LO	0.56	PV	0.56
5.5-6.0m deep	m	LO	0.60	PV	0.60
Steel pipes with electric resistant welded joints 200mm nominal bore					
not in trenches	m	LL	0.16	PU	0.16
not exceeding 1.5m deep	m	LL	0.28	PU	0.28
1.5-2.0m deep	m	LL	0.32	PU	0.32
2.0-2.5m deep	m	LL	0.36	PU	0.36
2.5-3.0m deep	m	LL	0.40	PU	0.40
3.0-3.5m deep	m	LL	0.44	PU	0.44
3.5-4.0m deep	m	LO	0.48	PV	0.48
4.0-4.5m deep	m	LO	0.52	PV	0.52
4.5-5.0m deep	m	LO	0.56	PV	0.56
5.0-5.5m deep	m	LO	0.60	PV	0.60
5.5-6.0m deep	m	LO	0.64	PV	0.64

	Unit	Labour grade	Labour hours	Plant grade	Plant hours
Steel pipes with electric resistant welded joints 250mm nominal bore					
not in trenches	m	LL	0.20	PU	0.20
not exceeding 1.5m deep	m	LL	0.32	PU	0.32
1.5-2.0m deep	m	LL	0.36	PU	0.36
2.0-2.5m deep	m	LL	0.40	PU	0.40
2.5-3.0m deep	m	LL	0.44	PU	0.44
3.0-3.5m deep	m	LL	0.48	PU	0.48
3.5-4.0m deep	m	LO	0.52	PV	0.52
4.0-4.5m deep	m	LO	0.56	PV	0.56
4.5-5.0m deep	m	LO	0.60	PV	0.60
5.0-5.5m deep	m	LO	0.64	PV	0.64
5.5-6.0m deep	m	LO	0.68	PV	0.68
Steel pipes with electric resistant welded joints 300mm nominal bore					
not in trenches	m	LL	0.24	PU	0.24
not exceeding 1.5m deep	m	LL	0.36	PU	0.36
1.5-2.0m deep	m	LL	0.40	PU	0.40
2.0-2.5m deep	m	LL	0.44	PU	0.44
2.5-3.0m deep	m	LL	0.48	PU	0.48
3.0-3.5m deep	m	LL	0.52	PU	0.52
3.5-4.0m deep	m	LO	0.56	PV	0.56
4.0-4.5m deep	m	LO	0.60	PV	0.60
4.5-5.0m deep	m	LO	0.64	PV	0.64
5.0-5.5m deep	m	LO	0.68	PV	0.68
5.5-6.0m deep	m	LO	0.72	PV	0.72

	Unit	Labour grade	Labour hours	Plant grade	Plant hours
PVC-U pipes with ring seal joints, 82mm nominal bore					
not in trenches	m	LL	0.06	PU	0.06
not exceeding 1.5m deep	m	LL	0.10	PU	0.10
1.5-2.0m deep	m	LL	0.13	PU	0.13
2.0-2.5m deep	m	LL	0.16	PU	0.16
2.5-3.0m deep	m	LL	0.19	PU	0.19
3.0-3.5m deep	m	LL	0.22	PU	0.22
3.5-4.0m deep	m	LO	0.25	PV	0.25
4.0-4.5m deep	m	LO	0.28	PV	0.28
4.5-5.0m deep	m	LO	0.31	PV	0.31
5.0-5.5m deep	m	LO	0.34	PV	0.34
5.5-6.0m deep	m	LO	0.37	PV	0.37
PVC-U pipes with ring seal joints, 110mm nominal bore					
not in trenches	m	LL	0.08	PU	0.08
not exceeding 1.5m deep	m	LL	0.13	PU	0.13
1.5-2.0m deep	m	LL	0.16	PU	0.16
2.0-2.5m deep	m	LL	0.19	PU	0.19
2.5-3.0m deep	m	LL	0.22	PU	0.22
3.0-3.5m deep	m	LL	0.25	PU	0.25
3.5-4.0m deep	m	LO	0.28	PV	0.28
4.0-4.5m deep	m	LO	0.31	PV	0.31
4.5-5.0m deep	m	LO	0.34	PV	0.34
5.0-5.5m deep	m	LO	0.37	PV	0.37
5.5-6.0m deep	m	LO	0.40	PV	0.40

	Unit	Labour grade	Labour hours	Plant grade	Plant hours
PVC-U pipes with ring seal joints, 160mm nominal bore					
not in trenches	m	LL	0.10	PU	0.10
not exceeding 1.5m deep	m	LL	0.16	PU	0.16
1.5-2.0m deep	m	LL	0.19	PU	0.19
2.0-2.5m deep	m	LL	0.22	PU	0.22
2.5-3.0m deep	m	LL	0.25	PU	0.25
3.0-3.5m deep	m	LL	0.28	PU	0.28
3.5-4.0m deep	m	LO	0.31	PV	0.31
4.0-4.5m deep	m	LO	0.34	PV	0.34
4.5-5.0m deep	m	LO	0.37	PV	0.37
5.0-5.5m deep	m	LO	0.40	PV	0.40
5.5-6.0m deep	m	LO	0.43	PV	0.43

Pipe fittings

Vitrified clay bends, nominal bore

	Unit	Labour grade	Labour hours	Plant grade	Plant hours
100mm	nr	LL	0.04	-	-
150mm	nr	LL	0.07	-	-
225mm	nr	LL	0.09	-	-
300mm	nr	LL	0.14	-	-
400mm	nr	LL	0.20	-	-
450mm	nr	LL	0.25	-	-

Vitrified clay bends, nominal bore

	Unit	Labour grade	Labour hours	Plant grade	Plant hours
100mm	nr	LL	0.04	-	-
150mm	nr	LL	0.07	-	-
225mm	nr	LL	0.09	-	-
300mm	nr	LL	0.14	-	-
400mm	nr	LL	0.20	-	-
450mm	nr	LL	0.25	-	-

	Unit	Labour grade	Labour hours	Plant grade	Plant hours
Vitrified clay rest bends, nominal bore					
100mm	nr	LL	0.05	-	-
150mm	nr	LL	0.08	-	-
225mm	nr	LL	0.10	-	-
300mm	nr	LL	0.15	-	-
Vitrified clay single junctions, nominal bore					
100mm	nr	LL	0.08	-	-
150mm	nr	LL	0.10	-	-
225mm	nr	LL	0.17	-	-
400mm	nr	LL	0.25	-	-
475mm	nr	LL	0.50	-	-
Vitrified clay double junctions, nominal bore					
100mm	nr	LL	0.20	-	-
150mm	nr	LL	0.30	-	-
225mm	nr	LL	0.35	-	-
300mm	nr	LL	0.40	-	-
Vitrified clay tapers, nominal bore					
150mm	nr	LL	0.04	-	-
225mm	nr	LL	0.07	-	-
300mm	nr	LL	0.09	-	-
375mm	nr	LL	0.14	-	-
Vitrified clay saddles, nominal bore					
100mm	nr	LL	0.50	-	-
150mm	nr	LL	1.00	-	-
225mm	nr	LL	1.50	-	-
300mm	nr	LL	2.00	-	-

	Unit	Labour grade	Labour hours	Plant grade	Plant hours
Concrete bends					
300mm	nr	LL	0.03	PV	0.07
375mm	nr	LL	0.05	PV	0.09
450mm	nr	LL	0.08	PV	0.12
525mm	nr	LL	0.12	PV	0.15
600mm	nr	LL	0.15	PV	0.17
750mm	nr	LO	0.17	PU	0.17
900mm	nr	LO	0.25	PU	0.30
1200mm	nr	LO	0.40	PU	0.45
1800mm	nr	LO	0.45	PU	0.60
2100mm	nr	LO	0.50	PU	0.70
Concrete single junction, tumbling bay, nominal bore					
300mm	nr	LL	0.05	PV	0.10
375mm	nr	LL	0.07	PV	0.13
450mm	nr	LL	0.12	PV	0.18
525mm	nr	LL	0.18	PV	0.22
600mm	nr	LL	0.22	PV	0.35
Concrete double junction, tumbling bay, nominal bore					
300mm	nr	LL	0.06	PV	0.14
375mm	nr	LL	0.10	PV	0.18
Ductile iron all-socket bends, nominal bore					
80mm	nr	LL	0.30	-	-
100mm	nr	LL	0.35	-	-
150mm	nr	LL	0.20	PV	0.20
250mm	nr	LL	0.25	PV	0.25
300mm	nr	LL	0.30	PV	0.30

	Unit	Labour grade	Labour hours	Plant grade	Plant hours
Ductile iron all-socket tees, main nominal bore					
80mm	nr	LL	0.45	-	-
100mm	nr	LL	0.50	-	-
150mm	nr	LL	0.55	-	-
200mm	nr	LL	0.30	PV	0.30
250mm	nr	LL	0.35	PV	0.35
300mm	m	LL	0.40	PV	0.40
Ductile iron all-socket angle branch, main nominal bore					
80mm	nr	LL	0.45	-	-
100mm	nr	LL	0.50	-	-
150mm	nr	LL	0.55	-	-
200mm	nr	LL	0.30	PV	0.30
250mm	nr	LL	0.35	PV	0.35
300mm	nr	LL	0.40	PV	0.40
Ductile iron flanged bends, nominal bore					
80mm	nr	LL	0.45	-	-
100mm	nr	LL	0.50	-	-
150mm	nr	LL	0.55	-	-
200mm	nr	LL	0.30	PV	0.30
250mm	nr	LL	0.40	PV	0.50
Ductile iron flanged tees, main nominal bore					
80mm	nr	LL	0.55	-	-
100mm	nr	LL	0.60	-	-
150mm	nr	LL	0.65	-	-
200mm	nr	LL	0.40	PV	0.40
250mm	nr	LL	0.45	PV	0.45
300mm	nr	LL	0.50	PV	0.50

	Unit	Labour grade	Labour hours	Plant grade	Plant hours
Ductile iron flanged angle branch, main nominal bore					
80mm	nr	LL	0.55	-	-
100mm	nr	LL	0.60	-	-
150mm	nr	LL	0.65	-	-
200mm	nr	LL	0.40	PV	0.95
250mm	nr	LL	0.45	PV	1.25
300mm	nr	LL	0.50	PV	1.40
Ductile iron flanged bellmouth, main nominal bore					
80mm	nr	LL	0.45	-	-
100mm	nr	LL	0.50	-	-
150mm	nr	LL	0.55	-	-
200mm	nr	LL	0.30	PV	0.18
250mm	nr	LL	0.35	PV	0.20
300mm	nr	LL	0.40	PV	0.22

Cast iron valves and penstocks, nominal bore

Flanged gate valves, hand operated with handwheel, nominal bore

	Unit	Labour grade	Labour hours	Plant grade	Plant hours
100mm	nr	LL	0.20	PV	0.20
150mm	nr	LL	0.30	PV	0.30
200mm	nr	LL	0.40	PV	0.40
250mm	nr	LL	0.50	PV	0.50
300mm	nr	LL	0.60	PV	0.60

	Unit	Labour grade	Labour hours	Plant grade	Plant hours

Circular pattern penstock, hand operated with handwheel, nominal bore

	Unit	Labour grade	Labour hours	Plant grade	Plant hours
300mm	nr	LL	1.00	PV	1.00
350mm	nr	LL	1.60	PV	1.60
450mm	nr	LL	2.10	PV	2.10
500mm	nr	LL	2.20	PV	2.20
600mm	nr	LL	2.50	PV	2.50
700mm	nr	LL	2.70	PV	2.70
750mm	nr	LL	3.20	PV	3.20

	Unit	Labour grade	Labour hours	Plant grade	Plant hours	Filling m3

Pipe beds

Sand or granular material bed 150mm thick to pipe, nominal bore

	Unit	Labour grade	Labour hours	Plant grade	Plant hours	Filling m3
100mm	m	LL	0.04	PV	0.04	0.068
150mm	m	LL	0.04	PV	0.04	0.079
225mm	m	LL	0.05	PV	0.05	0.090
300mm	m	LL	0.06	PV	0.06	0.113
400mm	m	LL	0.07	PV	0.07	0.120
450mm	m	LL	0.08	PV	0.08	0.128
525mm	m	LL	0.09	PV	0.09	0.138
600mm	m	LO	0.07	PU	0.07	0.155
675mm	m	LO	0.07	PU	0.07	0.170
750mm	m	LO	0.08	PU	0.08	0.180
900mm	m	LO	0.09	PU	0.09	0.206
1200mm	m	LO	0.13	PU	0.13	0.259
1500mm	m	LO	0.14	PU	0.14	0.311
1800mm	m	LO	0.16	PU	0.16	0.360
2100mm	m	LO	0.18	PU	0.18	0.400

	Unit	Labour grade	Labour hours	Plant grade	Plant hours	Filling m3

Concrete in bed 150mm
thick to pipe, nominal bore

	Unit	Labour grade	Labour hours	Plant grade	Plant hours	Filling m3
100mm	m	LL	0.08	PV	0.08	0.068
150mm	m	LL	0.08	PV	0.08	0.079
225mm	m	LL	0.10	PV	0.10	0.090
300mm	m	LL	0.12	PV	0.12	0.113
400mm	m	LL	0.14	PV	0.14	0.120
450mm	m	LL	0.15	PV	0.15	0.128
525mm	m	LL	0.16	PV	0.16	0.138
600mm	m	LO	0.17	PU	0.17	0.155
675mm	m	LO	0.18	PU	0.18	0.180
750mm	m	LO	0.19	PU	0.19	0.170
900mm	m	LO	0.20	PU	0.20	0.206
1200mm	m	LO	0.26	PU	0.26	0.259
1500mm	m	LO	0.28	PU	0.28	0.311
1800mm	m	LO	0.32	PU	0.32	0.360
2100mm	m	LO	0.36	PU	0.36	0.400

Sand or granular
material in bed 150mm
thick and haunching
to pipe, nominal bore

	Unit	Labour grade	Labour hours	Plant grade	Plant hours	Filling m3
100mm	m	LL	0.06	PV	0.06	0.117
150mm	m	LL	0.06	PV	0.06	0.152
225mm	m	LL	0.08	PV	0.08	0.195
300mm	m	LL	0.08	PV	0.08	0.279
400mm	m	LL	0.10	PV	0.10	0.285
450mm	m	LL	0.10	PV	0.10	0.315
525mm	m	LL	0.12	PV	0.12	0.330
600mm	m	LO	0.12	PU	0.12	0.346
675mm	m	LO	0.12	PU	0.12	0.380
750mm	m	LO	0.13	PU	0.13	0.427
900mm	m	LO	0.13	PU	0.13	0.528
1200mm	m	LO	0.18	PU	0.18	0.725
1500mm	m	LO	0.20	PU	0.20	0.953
1800mm	m	LO	0.24	PU	0.24	1.181
2100mm	m	LO	0.26	PU	0.26	1.409

	Unit	Labour grade	Labour hours	Plant grade	Plant hours	Filling m3

Concrete in bed 150mm
thick and haunching
to pipe, nominal bore

	Unit	Labour grade	Labour hours	Plant grade	Plant hours	Filling m3
100mm	m	LL	0.08	PV	0.08	0.117
150mm	m	LL	0.12	PV	0.12	0.152
225mm	m	LL	0.15	PV	0.15	0.195
300mm	m	LL	0.20	PV	0.20	0.279
400mm	m	LL	0.22	PV	0.22	0.285
450mm	m	LL	0.23	PV	0.23	0.315
525mm	m	LL	0.24	PV	0.24	0.330
600mm	m	LO	0.30	PU	0.30	0.346
675mm	m	LO	0.22	PU	0.22	0.380
750mm	m	LO	0.25	PU	0.25	0.427
900mm	m	LO	0.28	PU	0.28	0.528
1200mm	m	LO	0.30	PU	0.30	0.725
1500mm	m	LO	0.32	PU	0.32	0.953
1800mm	m	LO	0.34	PU	0.34	1.181
2100mm	m	LO	0.36	PU	0.36	1.409

Sand or granular material
in bed 150mm
thick and surround
to pipe, nominal bore

	Unit	Labour grade	Labour hours	Plant grade	Plant hours	Filling m3
100mm	m	LL	0.05	PV	0.05	0.185
150mm	m	LL	0.07	PV	0.07	0.231
225mm	m	LL	0.08	PV	0.08	0.285
300mm	m	LL	0.10	PV	0.10	0.391
400mm	m	LL	0.12	PV	0.12	0.438
450mm	m	LL	0.14	PV	0.14	0.483
525mm	m	LL	0.16	PV	0.16	0.557
600mm	m	LO	0.18	PU	0.18	0.635
675mm	m	LO	0.18	PU	0.18	0.706
750mm	m	LO	0.19	PU	0.19	0.791
900mm	m	LO	0.19	PU	0.19	0.960
1200mm	m	LO	0.20	PU	0.20	1.361
1500mm	m	LO	0.21	PU	0.21	1.798
1800mm	m	LO	0.22	PU	0.22	2.228
2100mm	m	LO	0.24	PU	0.24	3.088

	Unit	Labour grade	Labour hours	Plant grade	Plant hours	Filling m3

Concrete in bed 150mm
thick and surround
to pipe, nominal bore

	Unit	Labour grade	Labour hours	Plant grade	Plant hours	Filling m3
100mm	m	LL	0.07	PV	0.07	0.185
150mm	m	LL	0.10	PV	0.10	0.231
225mm	m	LL	0.12	PV	0.12	0.285
300mm	m	LL	0.15	PV	0.15	0.391
400mm	m	LL	0.17	PV	0.17	0.438
450mm	m	LL	0.19	PV	0.19	0.483
525mm	m	LL	0.22	PV	0.22	0.557
600mm	m	LO	0.22	PU	0.22	0.635
675mm	m	LO	0.24	PU	0.24	0.706
750mm	m	LO	0.25	PU	0.25	0.791
900mm	m	LO	0.26	PU	0.26	0.960
1200mm	m	LO	0.28	PU	0.28	1.361
1500mm	m	LO	0.29	PU	0.29	1.798
1800mm	m	LO	0.30	PU	0.30	2.228
2100mm	m	LO	0.32	PU	0.32	3.088

	Unit	Labour grade	Labour hours	Plant grade	Plant hours

Manholes

Excavation for manholes
including backfilling,
earthwork support and
disposal of excavated
material

	Unit	Labour grade	Labour hours	Plant grade	Plant hours
1-2m deep	m	LK	0.30	PW	0.04
2-5m deep	m	LK	0.40	PW	0.07

Concrete in bases, thickness

	Unit	Labour grade	Labour hours	Plant grade	Plant hours
not exceeding 150mm	m3	LK	0.24	PW	0.18
150-300mm	m3	LK	0.22	PW	0.16

	Unit	Labour grade	Labour hours	Plant grade	Plant hours
Concrete in surrounds, thickness					
not exceeding 150mm	m3	LK	0.30	PW	0.30
150-300mm	m3	LK	0.26	PW	0.26
Concrete in bottom of manhole, thickness					
not exceeding 150mm	m3	LK	1.00	PW	0.90
150-300mm	m3	LK	0.90	PW	1.80

Precast concrete units

Shaft rings, diameter

675mm	m	LK	0.60	PW	0.60
900mm	m	LK	0.70	PW	0.70
1050mm	m	LK	0.80	PW	0.80
1200mm	m	LK	0.90	PW	0.90
1350mm	m	LK	1.00	PW	1.00
1500mm	m	LK	1.10	PW	1.10
1800mm	m	LK	1.20	PW	1.20

Cover slabs, diameter

675mm	m	LK	0.40	PW	0.60
900mm	m	LK	0.45	PW	0.70
1050mm	m	LK	0.50	PW	0.50
1200mm	m	LK	0.55	PW	0.55
1350mm	m	LK	0.60	PW	0.60
1500mm	m	LK	0.65	PW	0.65
1800mm	m	LK	0.70	PW	0.70
1350 x 1125mm	m	LK	0.45	PW	0.45
1650 x 1500mm	m	LK	0.55	PW	0.55

Tapers, diameter

1200-900mm	m	LK	0.55	PW	0.55
1350-900mm	m	LK	0.60	PW	0.60
1500-900mm	m	LK	0.65	PW	0.65
1800-900mm	m	LK	0.70	PW	0.70

	Unit	Labour grade	Labour hours	Plant grade	Plant hours
Engineering bricks in cement mortar (1:3)					
Half brick wall	m2	LK	0.34	PW	0.34
One brick wall	m2	LK	0.60	PW	0.60
Vitrified clayware half round channels, bedded and pointed in cement mortar (1:3)					
100mm	m	LK	0.12	PW	0.12
150mm	m	LK	0.14	PW	0.14
225mm	m	LK	0.16	PW	0.16
300mm	m	LK	0.20	PW	0.20
Vitrified clayware half round bends, bedded and pointed in cement mortar (1:3)					
100mm	nr	LK	0.10	PW	0.10
150mm	nr	LK	0.12	PW	0.12
225mm	nr	LK	0.14	PW	0.14
300mm	nr	LK	0.16	PW	0.16
Vitrified clayware half round tapers, bedded and pointed in cement mortar (1:3)					
150-100mm	nr	LK	0.10	PW	0.10
225-150mm	nr	LK	0.12	PW	0.12
300-225mm	nr	LK	0.14	PW	0.14
Vitrified clayware three-quarter section branch channel bends, bedded and pointed in cement mortar (1:3)					
100mm	nr	LK	0.12	PW	0.12
150mm	nr	LK	0.14	PW	0.14
225mm	nr	LK	0.16	PW	0.16
300mm	nr	LK	0.20	PW	0.20

	Unit	Labour grade	Labour hours	Plant grade	Plant hours
Galvanised metal step irons built into engineering brickwork	nr	LK	0.05	PW	0.05
Ductile iron manholes and frames Grade C, light duty, size 600 x 450mm bedded in cement mortar (1:3), bedded in sand and grease	nr	LK	0.25	PW	0.25
Ductile iron manholes and frames Grade C, light duty, size 600 x 600mm bedded in cement mortar (1:3), bedded in sand and grease	nr	LK	0.25	PW	0.25
Ductile iron manholes and frames Grade B, medium duty, size 600 x 450mm bedded in cement mortar (1:3), bedded in sand and grease	nr	LK	0.30	PW	0.30
Ductile iron manholes and frames Grade B, medium duty, size 600 x 600mm bedded in cement mortar (1:3), bedded in sand and grease	nr	LK	0.30	PW	0.30
Ductile iron manholes and frames Grade B, medium duty, size 550mm diameter bedded in cement mortar (1:3), bedded in sand and grease	nr	LK	0.30	PW	0.30

	Unit	Labour grade	Labour hours	Plant grade	Plant hours
Ductile iron manholes and frames Grade A, heavy duty, size 550 x 500mm bedded in cement mortar (1:3), bedded in sand and grease	nr	LK	0.30	PW	0.30
Ductile iron manholes and frames Grade A, heavy duty, size 600 x 500mm bedded in cement mortar (1:3), bedded in sand and grease	nr	LK	0.30	PW	0.30

French drains

	Unit	Labour grade	Labour hours	Plant grade	Plant hours
French drains, rubble drains, ditches and trenches filled with granular material					
20mm aggregate	m3	LK	0.15	-	-
40mm aggregate	m3	LK	0.15	-	-
French drains, rubble drains, ditches and trenches filled with brick rubble	m3	LK	0.15	-	-
Trenches for unpiped rubble drains, cross-sectional area					
not exceeding 0.25m2	m	LL	0.02	PU	0.03
0.25-0.50m2	m	LL	0.07	PU	0.10
0.50-0.75m2	m	LL	0.10	PU	0.13
0.75-1.00m2	m	LL	0.13	PU	0.15
1.00-1.50m2	m	LL	0.16	PU	0.18
1.50-2.00m2	m	LL	0.20	PU	0.25
2.00-3.00m2	m	LL	0.25	PU	0.36
4.00m2	m	LL	0.30	PU	0.43

	Unit	Labour grade	Labour hours	Plant grade	Plant hours

Ditches

Rectangular section
diches, unlined, cross-
sectional area

	Unit	Labour grade	Labour hours	Plant grade	Plant hours
not exceeding 0.25m2	m	LL	0.03	PU	0.04
0.25-0.50m2	m	LL	0.08	PU	0.11
0.50-0.75m2	m	LL	0.11	PU	0.14
0.75-1.00m2	m	LL	0.14	PU	0.16
1.00-1.50m2	m	LL	0.18	PU	0.20
1.50-2.00m2	m	LL	0.22	PU	0.28
2.00-3.00m2	m	LL	0.28	PU	0.34
4m2	m	LL	0.33	PU	0.35

Rectangular section
ditches, lined with
concrete 100mm thick,
cross-sectional area

	Unit	Labour grade	Labour hours	Plant grade	Plant hours
not exceeding 0.25m2	m	LL	0.06	PU	0.08
0.25-0.50m2	m	LL	0.12	PU	0.15
0.50-0.75m2	m	LL	0.16	PU	0.21
0.75-1.00m2	m	LL	0.20	PU	0.25
1.00-1.50m2	m	LL	0.16	PU	0.32
1.50-2.00m2	m	LL	0.32	PU	0.56
2.00-3.00m2	m	LL	0.42	PU	0.56
4m2	m	LL	0.51	PU	0.62

Vee section ditches,
unlined, cross-
sectional area

	Unit	Labour grade	Labour hours	Plant grade	Plant hours
not exceeding 0.25m2	m	LL	0.01	PU	0.02
0.25-0.50m2	m	LL	0.06	PU	0.08
0.50-0.75m2	m	LL	0.09	PU	0.11
0.75-1.00m2	m	LL	0.11	PU	0.13
1.00-1.50m2	m	LL	0.14	PU	0.16
1.50-2.00m2	m	LL	0.18	PU	0.22
2.00-3.00m2	m	LL	0.22	PU	0.27
4m2	m	LL	0.25	PU	0.30

	Unit	Labour grade	Labour hours	Plant grade	Plant hours
Rectangular section diches, unlined, cross-sectional area					
not exceeding 0.25m2	m	LL	0.03	PU	0.04
0.25-0.50m2	m	LL	0.08	PU	0.11
0.50-0.75m2	m	LL	0.11	PU	0.14
0.75-1.00m2	m	LL	0.14	PU	0.16
1.00-1.50m2	m	LL	0.18	PU	0.20
1.50-2.00m2	m	LL	0.22	PU	0.28
2.00-3.00m2	m	LL	0.28	PU	0.34
4m2	m	LL	0.33	PU	0.35
Vee section ditches, lined with concrete 100mm thick, cross-sectional area					
not exceeding 0.25m2	m	LL	0.05	PU	0.03
0.25-0.50m2	m	LL	0.14	PU	0.10
0.50-0.75m2	m	LL	0.20	PU	0.15
0.75-1.00m2	m	LL	0.25	PU	0.19
1.00-1.50m2	m	LL	0.34	PU	0.26
1.50-2.00m2	m	LL	0.46	PU	0.34
2.00-3.00m2	m	LL	0.63	PU	0.46
4m2	m	LL	0.78	PU	0.57

Cable ducts

Vitrified clay one-way nominal bore 100mm, trench depth	Unit	Labour grade	Labour hours	Plant grade	Plant hours
not exceeding 1.5m	m	LL	0.14	PU	0.14
1.5-2m	m	LL	0.18	PU	0.18
2-2.5m	m	LL	0.24	PU	0.24
2.5-3m	m	LL	0.29	PU	0.29
3-3.5m	m	LL	0.36	PU	0.36
3.5-4m	m	LL	0.49	PU	0.49

	Unit	Labour grade	Labour hours	Plant grade	Plant hours
Vitrified clay one-way nominal bore 125mm, trench depth					
not exceeding 1.5m	m	LL	0.14	PU	0.14
1.5-2m	m	LL	0.18	PU	0.18
2-2.5m	m	LL	0.24	PU	0.24
2.5-3m	m	LL	0.29	PU	0.29
3-3.5m	m	LL	0.36	PU	0.36
3.5-4m	m	LL	0.49	PU	0.49
Vitrified clay one-way nominal bore 150mm, trench depth					
not exceeding 1.5m	m	LL	0.17	PU	0.14
1.5-2m	m	LL	0.23	PU	0.18
2-2.5m	m	LL	0.27	PU	0.24
2.5-3m	m	LL	0.33	PU	0.29
3-3.5m	m	LL	0.40	PU	0.36
3.5-4m	m	LL	0.54	PU	0.54
Vitrified clay one-way nominal bore 225mm, trench depth					
not exceeding 1.5m	m	LL	0.23	PU	0.23
1.5-2m	m	LL	0.29	PU	0.29
2-2.5m	m	LL	0.33	PU	0.33
2.5-3m	m	LL	0.39	PU	0.39
3-3.5m	m	LL	0.47	PU	0.47
3.5-4m	m	LL	0.60	PU	0.60

	Unit	Labour grade	Labour hours	Plant grade	Plant hours

Culverts

Galvanised bitumen-coated
sectional corrugated
culvert 1.5mm thick,
0.5m diameter, trench
depth

	Unit	Labour grade	Labour hours	Plant grade	Plant hours
not exceeding 1.5m	m	LL	0.08	PU	0.08
1.5-2m	m	LL	0.18	PU	0.18
2-2.5m	m	LL	0.20	PU	0.20
2.5-3m	m	LL	0.30	PU	0.30
3-3.5m	m	LL	0.35	PU	0.35
3.5-4m	m	LL	0.40	PU	0.40

Galvanised bitumen-coated
sectional corrugated
culvert 1.5mm thick,
1m diameter, trench
depth

	Unit	Labour grade	Labour hours	Plant grade	Plant hours
not exceeding 1.5m	m	LL	0.20	PU	0.20
1.5-2m	m	LL	0.30	PU	0.30
2-2.5m	m	LL	0.33	PU	0.33
2.5-3m	m	LL	0.37	PU	0.37
3-3.5m	m	LL	0.40	PU	0.40
3.5-4m	m	LL	0.45	PU	0.45

Galvanised bitumen-coated
sectional corrugated
culvert 1.5mm thick,
1.5m diameter, trench
depth

	Unit	Labour grade	Labour hours	Plant grade	Plant hours
not exceeding 1.5m	m	LL	0.30	PU	0.30
1.5-2m	m	LL	0.34	PU	0.34
2-2.5m	m	LL	0.40	PU	0.40
2.5-3m	m	LL	0.50	PU	0.50
3-3.5m	m	LL	0.60	PU	0.60
3.5-4m	m	LL	0.70	PU	0.70

	Unit	Labour grade	Labour hours	Plant grade	Plant hours
Galvanised bitumen-coated sectional corrugated culvert 1.5mm thick, 2m diameter, trench depth					
not exceeding 1.5m	m	LL	0.32	PU	0.32
1.5-2m	m	LL	0.37	PU	0.37
2-2.5m	m	LL	0.45	PU	0.45
2.5-3m	m	LL	0.60	PU	0.60
3-3.5m	m	LL	0.70	PU	0.70
3.5-4m	m	LL	0.80	PU	0.80

25

Roads

Weights of materials	kg/m3
Ashes	800
Asphalt	2750
Clinker	1450
Granite	2700
Gravel	1750
Hardcore	1900
Limestone	2350
Sandstone	2500
Slag	2800
Whinstone	2800
Precast concrete kerb	kg/m
175 x 50mm	22
250 x 150mm	88
300 x 150mm	100

Coverage areas of surface dressing

	m2/tonne
Aggregate/gravel	
3mm	160
6mm	140
9mm	120
14mm	90
20mm	75
Sand (per m3)	170

Labour grades

1 Unskilled operative	LC
1 Craftsman and 2 unskilled labourers	LJ
1 Craftsman and 1 unskilled labourer	LK
1 Ganger, 1 skilled operative and 1 unskilled operative.	LN
1 Ganger, 2 skilled operatives and 1 unskilled operative	LO

Plant grades

1 Crawler crane	PS
1 Crawler tractor and 1 motorised roller	PX
1 Concrete paver and 1 motorised roller	PY

	Unit	Labour grade	Labour hours	Plant grade	Plant hours	Materials tonnes

Sub-bases, flexible road bases and surfaces

Granular material DTp specified type 1, depth

	Unit	Labour grade	Labour hours	Plant grade	Plant hours	Materials tonnes
100mm	m2	LN	0.03	PX	0.02	0.189
150mm	m2	LN	0.03	PX	0.02	0.284
200mm	m2	LN	0.03	PX	0.02	0.378
300mm	m2	LN	0.04	PX	0.03	0.567
400mm	m2	LN	0.05	PX	0.03	0.756
500mm	m2	LN	0.05	PX	0.03	0.945

Granular material DTp specified type 2, depth

	Unit	Labour grade	Labour hours	Plant grade	Plant hours	Materials tonnes
100mm	m2	LN	0.03	PX	0.02	0.189
150mm	m2	LN	0.03	PX	0.02	0.284
200mm	m2	LN	0.03	PX	0.02	0.378
300mm	m2	LN	0.04	PX	0.03	0.567
400mm	m2	LN	0.05	PX	0.03	0.756
500mm	m2	LN	0.05	PX	0.03	0.945

Soil cement (100kg cement/m3 soil), depth

	Unit	Labour grade	Labour hours	Plant grade	Plant hours	Materials tonnes
100mm	m2	LN	0.03	PX	0.02	0.147
150mm	m2	LN	0.03	PX	0.02	0.221
200mm	m2	LN	0.03	PX	0.02	0.294
300mm	m2	LN	0.04	PX	0.03	0.441
400mm	m2	LN	0.05	PX	0.03	0.588
500mm	m2	LN	0.05	PX	0.03	0.735

	Unit	Labour grade	Labour hours	Plant grade	Plant hours	Materials tonnes
Cement bound aggregate (100kg cement/m3 soil), depth						
100mm	m2	LN	0.03	PX	0.02	0.185
150mm	m2	LN	0.03	PX	0.02	0.278
200mm	m2	LN	0.03	PX	0.02	0.370
300mm	m2	LN	0.04	PX	0.03	0.555
400mm	m2	LN	0.05	PX	0.03	0.740
500mm	m2	LN	0.05	PX	0.03	0.925
Lean concrete DTp specified strength 10, depth						
100mm	m2	LN	0.03	PX	0.02	0.100
150mm	m2	LN	0.03	PX	0.02	0.150
200mm	m2	LN	0.03	PX	0.02	0.200
300mm	m2	LN	0.04	PX	0.03	0.300
400mm	m2	LN	0.05	PX	0.03	0.400
500mm	m2	LN	0.05	PX	0.03	0.500
Hardcore,depth						
100mm	m2	LN	0.03	PX	0.02	0.100
150mm	m2	LN	0.03	PX	0.02	0.150
200mm	m2	LN	0.03	PX	0.02	0.200
300mm	m2	LN	0.04	PX	0.03	0.300
400mm	m2	LN	0.05	PX	0.03	0.400
500mm	m2	LN	0.05	PX	0.03	0.500
Wet mix macadam, depth						
75mm	m2	LN	0.04	PX	0.04	0.187
100mm	m2	LN	0.04	PX	0.04	0.250
150mm	m2	LN	0.05	PX	0.05	0.375
200mm	m2	LN	0.05	PX	0.05	0.500

	Unit	Labour grade	Labour hours	Plant grade	Plant hours	Materials tonnes
Dry bound macadam, depth						
75mm	m2	LN	0.04	PX	0.04	0.173
100mm	m2	LN	0.04	PX	0.04	0.230
150mm	m2	LN	0.05	PX	0.05	0.345
200mm	m2	LN	0.05	PX	0.05	0.460
Dense bitumen macadam, depth						
40mm	m2	LN	0.04	PX	0.04	0.100
50mm	m2	LN	0.05	PX	0.05	0.125
75mm	m2	LN	0.06	PX	0.05	0.175
100mm	m2	LN	0.07	PX	0.07	0.250
Dense bitumen macadam, depth						
20mm	m2	LN	0.03	PX	0.03	0.050
30mm	m2	LN	0.03	PX	0.03	0.075
40mm	m2	LN	0.04	PX	0.04	0.100
50mm	m2	LN	0.05	PX	0.05	0.125
75mm	m2	LN	0.06	PX	0.05	0.175
100mm	m2	LN	0.07	PX	0.07	0.250
Open textured bitumen macadam, depth						
20mm	m2	LN	0.03	PX	0.03	0.045
30mm	m2	LN	0.03	PX	0.03	0.065
40mm	m2	LN	0.04	PX	0.04	0.090
50mm	m2	LN	0.05	PX	0.05	0.110
75mm	m2	LN	0.06	PX	0.05	0.165
100mm	m2	LN	0.07	PX	0.07	0.220

	Unit	Labour grade	Labour hours	Plant grade	Plant hours	Materials tonnes
Dense tar surfacing, depth						
30mm	m2	LN	0.03	PX	0.03	0.075
40mm	m2	LN	0.04	PX	0.04	0.100
50mm	m2	LN	0.05	PX	0.05	0.125
Cold asphalt, depth						
15mm	m2	LN	0.03	PX	0.03	0.045
25mm	m2	LN	0.04	PX	0.04	0.070
30mm	m2	LN	0.05	PX	0.05	0.090
Rolled asphalt, depth						
30mm	m2	LN	0.05	PX	0.05	0.090
50mm	m2	LN	0.07	PX	0.07	0.150
Coated chippings, nominal size						
8mm, 6kg per m2	m2	LN	0.02	PX	0.02	0.006
10mm, 6kg per m2	m2	LN	0.02	PX	0.02	0.006
12mm, 8kg per m2	m2	LN	0.03	PX	0.03	0.008
14mm, 10kg per m2	m2	LN	0.03	PX	0.03	0.010

Concrete pavings

Carriageway slab, depth

	Unit	Labour grade	Labour hours	Plant grade	Plant hours	Materials tonnes
100mm	m2	LN	0.01	PY	0.02	0.100
150mm	m2	LN	0.02	PY	0.03	0.150
225mm	m2	LN	0.02	PY	0.04	0.225
300mm	m2	LN	0.03	PY	0.07	0.300
350mm	m2	LN	0.04	PY	0.08	0.350
400mm	m2	LN	0.05	PY	0.09	0.400

	Unit	Labour grade	Labour hours	Plant grade	Plant hours	Materials tonnes

Steel fabric reinforcement, nominal size

	Unit	Labour grade	Labour hours	Plant grade	Plant hours	Materials tonnes
2.22kg/m2, A142	m2	LJ	0.06	PS	0.06	0.002
2.61kg/m2, C283	m2	LJ	0.06	PS	0.06	0.003
3.02kg/m2, A193	m2	LJ	0.06	PS	0.06	0.003
3.05kg/m2, B196	m2	LJ	0.06	PS	0.06	0.003
3.41kg/m2, C385	m2	LJ	0.06	PS	0.06	0.003
3.73kg/m2, B283	m2	LJ	0.06	PS	0.06	0.004
3.95kg/m2, A252	m2	LJ	0.06	PS	0.06	0.004
4.34kg/m2, C503	m2	LJ	0.06	PS	0.06	0.004
4.53kg/m2, B385	m2	LJ	0.06	PS	0.06	0.005
5.55kg/m2, C636	m2	LJ	0.06	PS	0.06	0.006
5.93kg/m2, B503	m2	LJ	0.06	PS	0.06	0.006

	Unit	Labour grade	Labour hours	Plant grade	Plant hours	Materials m

Plain round mild steel bar reinforcement, nominal size

	Unit	Labour grade	Labour hours	Plant grade	Plant hours	Materials m
6mm	t	LJ	9.00	PS	9.00	4500
8mm	t	LJ	8.50	PS	8.50	2500
10mm	t	LJ	8.50	PS	8.50	1500
12mm	t	LJ	7.00	PS	7.00	1100
16mm	t	LJ	7.00	PS	7.00	650
20mm	t	LJ	5.00	PS	5.00	400
25mm	t	LJ	5.00	PS	5.00	250
32mm	t	LJ	4.50	PS	4.50	150

Deformed high yield mild steel bar reinforcement, nominal size

	Unit	Labour grade	Labour hours	Plant grade	Plant hours	Materials m
6mm	t	LJ	9.00	PS	9.00	4500
8mm	t	LJ	8.50	PS	8.50	2500
10mm	t	LJ	8.50	PS	8.50	1500
12mm	t	LJ	7.00	PS	7.00	1100
16mm	t	LJ	7.00	PS	7.00	650
20mm	t	LJ	5.00	PS	5.00	400
25mm	t	LJ	5.00	PS	5.00	250

	Unit	Labour grade	Labour hours	Plant grade	Plant hours

Joints

Longitudinal joints,
10mm diameter mild
steel bars at 500mm
centres, depth

	Unit	Labour grade	Labour hours	Plant grade	Plant hours
150mm	m	LK	0.50	-	-
225mm	m	LK	0.50	-	-
300mm	m	LK	0.50	-	-

Expansion joints,
10mm diameter mild
steel bars at 500mm
centres, 13mm wide,
depth

	Unit	Labour grade	Labour hours	Plant grade	Plant hours
150mm	m	LK	0.70	-	-
225mm	m	LK	0.70	-	-
300mm	m	LK	0.70	-	-

Expansion joints,
10mm diameter mild
steel bars at 500mm
centres, 19mm wide,
depth

	Unit	Labour grade	Labour hours	Plant grade	Plant hours
150mm	m	LK	0.70	-	-
225mm	m	LK	0.70	-	-
300mm	m	LK	0.70	-	-

Expansion joints,
10mm diameter mild
steel bars at 500mm
centres, 25mm wide,
depth

	Unit	Labour grade	Labour hours	Plant grade	Plant hours
150mm	m	LK	0.70	-	-
225mm	m	LK	0.70	-	-
300mm	m	LK	0.70	-	-

	Unit	Labour grade	Labour hours	Plant grade	Plant hours
Contraction joints, 10mm diameter mild steel bars at 500mm centres, depth					
150mm	m	LK	0.60	-	-
225mm	m	LK	0.60	-	-
300mm	m	LK	0.60	-	-
Warping joints, 10mm diameter mild steel bars at 300mm centres, depth					
150mm	m	LK	0.70	-	-
225mm	m	LK	0.70	-	-
300mm	m	LK	0.70	-	-

	Unit	Labour grade	Labour hours	Concrete m3	Mortar m3
Precast concrete					
Precast concrete kerbs straight or curved, bedded, pointed and jointed in cement mortar to radius exceeding 12m, size					
125 x 150mm	m	LO	0.06	0.014	0.003
125 x 255mm	m	LO	0.08	0.029	0.003
150 x 305mm	m	LO	0.10	0.037	0.003
Precast concrete kerbs straight or curved, bedded, pointed and jointed in cement mortar to radius not exceeding 12m, size					
125 x 150mm	m	LO	0.10	0.014	0.003
125 x 255mm	m	LO	0.12	0.029	0.003
150 x 305mm	m	LO	0.14	0.037	0.003

	Unit	Labour grade	Labour hours	Concrete m3	Mortar m3

Precast concrete kerbs straight or curved, bedded, pointed and jointed in cement mortar to radius not exceeding 5m, size

	Unit	Labour grade	Labour hours	Concrete m3	Mortar m3
125 x 150mm	m	LO	0.16	0.014	0.003
125 x 255mm	m	LO	0.18	0.029	0.003
150 x 305mm	m	LO	0.20	0.037	0.003

Quadrants, size

	Unit	Labour grade	Labour hours	Concrete m3	Mortar m3
305 x 305 x 150mm	nr	LO	0.06	0.009	0.001
305 x 305 x 255mm	nr	LO	0.07	0.009	0.001
455 x 455 x 150mm	nr	LO	0.10	0.009	0.001
455 x 455 x 255mm	nr	LO	0.12	0.009	0.001

Drop kerbs, size

	Unit	Labour grade	Labour hours	Concrete m3	Mortar m3
125 x 255mm	nr	LO	0.06	0.014	0.003
150 x 305mm	nr	LO	0.07	0.037	0.003

Precast concrete channels straight or curved, bedded, pointed and jointed in cement mortar to radius exceeding 12m, size

	Unit	Labour grade	Labour hours	Concrete m3	Mortar m3
125 x 150mm	m	LO	0.06	0.014	0.003
125 x 255mm	m	LO	0.08	0.029	0.003
150 x 305mm	m	LO	0.10	0.037	0.003

Precast concrete channels straight or curved, bedded, pointed and jointed in cement mortar to radius not exceeding 12m, size

	Unit	Labour grade	Labour hours	Concrete m3	Mortar m3
125 x 150mm	m	LO	0.10	0.014	0.003
125 x 255mm	m	LO	0.12	0.029	0.003
150 x 305mm	m	LO	0.14	0.037	0.003

	Unit	Labour grade	Labour hours	Concrete m3	Mortar m3

Precast concrete channels
straight or curved, bedded,
pointed and jointed in
cement mortar to radius
not exceeding 5m, size

	Unit	Labour grade	Labour hours	Concrete m3	Mortar m3
125 x 150mm	m	LO	0.16	0.014	0.003
125 x 255mm	m	LO	0.18	0.029	0.003
150 x 305mm	m	LO	0.20	0.037	0.003

26

Brickwork, blockwork and masonry

Weights of materials

Cement	1440kg/m3
Sand	1600kg/m3
Lime, ground	750kg/m3
Brickwork, 112.5mm	220kg/m2
215 mm	465kg/m2
327.5mm	710kg/m2
Stone, natural	2400kg/m3
reconstructed	2250kg/m3
Bricks, Fletton	1820kg/m2
engineering	2250kg/m2
concrete	1850kg/m2
Blocks, natural aggregate	
75mm thick	160kg/m2
100mm thick	215 kg/m2
140mm thick	300kg/m2

Blocks, lightweight aggregate

75mm thick	60kg/m2
100mm thick	80kg/m2
140mm thick	112kg/m2

Stone

Artificial	2200kg/m3
Bath	2200kg/m3
Darley Dale	2400kg/m3
Portland	2200kg/m3
York	2400kg/m3

Bricks per m2   **nr**

Brick size 215 x 103.5 x 65mm

Half brick wall

stretcher bond	59
English bond	89
English garden wall bond	74
Flemish bond	79

One brick wall

English bond	118
Flemish	118

One and a half brick wall

English bond	178
Flemish bond	178

Two brick wall

English bond	238
Flemish bond	238

Brick size 200 x 100 x 75mm

75mm thick	67
90mm thick	133
190mm thick	200

Brick size 200 x 100 x 75mm

90mm thick	50
190mm thick	100
290mm thick	150

Brick size 300 x 100 x 75mm

90mm thick	44

Brick size 300 x100 x 100mm

90mm thick	33

Blocks per m2   nr

Block size 414 x 215mm

60mm thick	9.9
75mm thick	9.9
100mm thick	9.9
140mm thick	9.9
190mm thick	9.9
215mm thick	9.9

Mortar per m2 in brick walling

Brick size 215 x 1035 x 65mm	Wirecut m3	1 frog m3	2 frogs m3
Half brick wall	0.017	0.024	0.031
One brick wall	0.045	0.059	0.073
One and a half brick wall	0.072	0.093	0.114
Two brick wall	0.101	0.128	0.155

Brick size 200 x 100 x 75mm	Solid m3	Perforated m3
90mm thick	0.016	0.019
190mm thick	0.042	0.048
290mm thick	0.068	0.078

Brick size 200 x 100 x 100mm	Solid m3	Perforated m3
90mm thick	0.013	0.016
190mm thick	0.036	0.041
290mm thick	0.059	0.067
Brick size 300 x 100 x 75mm		
90mm thick	0.015	0.018
Brick size 300 x 100 x 100mm		
90mm thick	0.015	-

Mortar per m2 in block walling

Block size 440 x 215mm

60mm thick	0.004	-
75mm thick	0.005	-
100mm	0.006	-
140mm	0.007	-
190mm	0.008	-
215mm	0.009	-

Mortar per m2 in random rubble walling

	m3
300mm thick	0.120
450mm thick	0.160
550mm thick	0.200

Length of pointing per m2 (one face only)

	m
English bond	19.1
English garden wall bond	18.1
Flemish bond	18.4
Flemish garden wall bond	17.7

Damp-proof courses

	kg/m2
Hessian base	3.8
Fibre base	3.3
Asbestos base	3.8
Hessian base and lead core	4.4
Asbestos base and lead core	4.9
Bitumen sheeting	5.4

Labour grades

Craftsman	LA
2 Bricklayers and 1 unskilled operative	LD
2 Masons and 1 unskilled labourer	LH

	Unit	Labour grade	Labour hours	Bricks nr	Mortar m3

Brickwork

Common bricks in cement mortar (1:3)

Half brick wall thick

	Unit	Labour grade	Labour hours	Bricks nr	Mortar m3
vertical straight walls	m2	LD	0.60	59	0.017
vertical curved walls	m2	LD	0.80	59	0.017
battered straight walls	m2	LD	0.80	59	0.017
battered curved walls	m2	LD	0.90	59	0.017
vertical backing	m2	LD	0.65	59	0.017
battered backing	m2	LD	0.75	59	0.017

One brick wall thick

	Unit	Labour grade	Labour hours	Bricks nr	Mortar m3
vertical straight walls	m2	LD	1.00	118	0.045
vertical curved walls	m2	LD	1.20	118	0.045
battered straight walls	m2	LD	1.20	118	0.045
battered curved walls	m2	LD	1.30	118	0.045
vertical backing	m2	LD	1.05	118	0.045
battered backing	m2	LD	1.15	118	0.045

One and a half brick wall thick

	Unit	Labour grade	Labour hours	Bricks nr	Mortar m3
vertical straight walls	m2	LD	1.40	178	0.072
vertical curved walls	m2	LD	1.60	178	0.072
battered straight walls	m2	LD	1.60	178	0.072
battered curved walls	m2	LD	1.70	178	0.072
vertical backing	m2	LD	1.45	178	0.072
battered backing	m2	LD	1.55	178	0.072

	Unit	Labour grade	Labour hours	Bricks nr	Mortar m3
Two brick wall thick					
vertical straight walls	m2	LD	2.00	238	0.101
vertical curved walls	m2	LD	2.20	238	0.101
battered straight walls	m2	LD	2.20	238	0.101
battered curved walls	m2	LD	2.30	238	0.101
vertical backing	m2	LD	2.05	238	0.101
battered backing	m2	LD	2.15	238	0.101
Surface features					
Brick on edge coping, pointing all round, width, 225mm	m	LD	0.10	13	0.005
Brick on end coping, pointing all round, width, 225mm	m	LD	0.10	26	0.010
Band course projecting 25mm from face of wall 112.5mm deep, pointing all round	m	LD	0.08	13	0.005
Fair face and flush pointing	m2	LD	0.15	-	-
Fair face and weather-struck pointing	m2	LD	0.18	-	-

	Unit	Labour grade	Labour hours	Bricks nr	Mortar m3
Facing bricks in cement mortar (1:3)					
Half brick wall thick					
vertical straight walls	m2	LD	0.80	59	0.017
vertical curved walls	m2	LD	1.00	59	0.017
battered straight walls	m2	LD	1.00	59	0.017
battered curved walls	m2	LD	1.10	59	0.017
vertical backing	m2	LD	0.85	59	0.017
battered backing	m2	LD	0.95	59	0.017
One brick wall thick					
vertical straight walls	m2	LD	1.20	118	0.045
vertical curved walls	m2	LD	1.40	118	0.045
battered straight walls	m2	LD	1.40	118	0.045
battered curved walls	m2	LD	1.50	118	0.045
vertical backing	m2	LD	1.25	118	0.045
battered backing	m2	LD	1.35	118	0.045
One and a half brick wall thick					
vertical straight walls	m2	LD	1.60	178	0.072
vertical curved walls	m2	LD	1.80	178	0.072
battered straight walls	m2	LD	1.80	178	0.072
battered curved walls	m2	LD	1.90	178	0.072
vertical backing	m2	LD	1.65	178	0.072
battered backing	m2	LD	1.75	178	0.072

	Unit	Labour grade	Labour hours	Bricks nr	Mortar m3
Two brick wall thick					
vertical straight walls	m2	LD	1.80	238	0.101
vertical curved walls	m2	LD	2.00	238	0.101
battered straight walls	m2	LD	2.00	238	0.101
battered curved walls	m2	LD	2.10	238	0.101
vertical backing	m2	LD	1.85	238	0.101
battered backing	m2	LD	1.95	238	0.101

Surface features

	Unit	Labour grade	Labour hours	Bricks nr	Mortar m3
Brick on edge coping, pointing all round, width, 225mm	m	LD	0.15	13	0.005
Brick on end coping, pointing all round, width, 225mm	m	LD	0.15	26	0.010
Band course projecting 25mm from face of wall 112.5mm deep, pointing all round	m	LD	0.10	13	0.005
Fair face and flush pointing	m2	LD	0.20	-	-
Fair face and weather-struck pointing	m2	LD	0.22	-	-

	Unit	Labour grade	Labour hours	Bricks nr	Mortar m3
Engineering bricks in cement mortar (1:3)					
Half brick wall thick					
vertical straight walls	m2	LD	0.70	59	0.017
vertical curved walls	m2	LD	0.90	59	0.017
battered straight walls	m2	LD	0.90	59	0.017
battered curved walls	m2	LD	1.00	59	0.017
vertical backing	m2	LD	0.75	59	0.017
battered backing	m2	LD	0.85	59	0.017
One brick wall thick					
vertical straight walls	m2	LD	1.10	118	0.045
vertical curved walls	m2	LD	1.30	118	0.045
battered straight walls	m2	LD	1.30	118	0.045
battered curved walls	m2	LD	1.40	118	0.045
vertical backing	m2	LD	1.15	118	0.045
battered backing	m2	LD	1.25	118	0.045
One and a half brick wall thick					
vertical straight walls	m2	LD	1.50	178	0.072
vertical curved walls	m2	LD	1.70	178	0.072
battered straight walls	m2	LD	1.70	178	0.072
battered curved walls	m2	LD	1.80	178	0.072
vertical backing	m2	LD	1.55	178	0.072
battered backing	m2	LD	1.65	178	0.072

	Unit	Labour grade	Labour hours	Bricks nr	Mortar m3
Two brick wall thick					
vertical straight walls	m2	LD	1.70	238	0.101
vertical curved walls	m2	LD	1.90	238	0.101
battered straight walls	m2	LD	1.90	238	0.101
battered curved walls	m2	LD	2.00	238	0.101
vertical backing	m2	LD	1.75	238	0.101
battered backing	m2	LD	1.85	238	0.101

Surface features

	Unit	Labour grade	Labour hours	Bricks nr	Mortar m3
Brick on edge coping, pointing all round, width, 225mm	m	LD	0.15	13	0.005
Brick on end coping, pointing all round, width, 225mm	m	LD	0.15	26	0.010
Band course projecting 25mm from face of wall 112.5mm deep, pointing all round	m	LD	0.10	13	0.005
Fair face and flush pointing	m2	LD	0.20	-	-
Fair face and weather-struck pointing	m2	LD	0.22	-	-

	Unit	Labour grade	Labour hours	Blocks nr	Mortar m3

Blockwork

Lightweight blockwork strength 3.5N/mm2, size 440 x 215mm in cement mortar (1:3)

100mm thick

vertical straight walls	m2	LD	0.40	9.9	0.006
vertical curved walls	m2	LD	0.50	9.9	0.006
battered straight walls	m2	LD	0.50	9.9	0.006
battered curved walls	m2	LD	0.65	9.9	0.006
vertical backing	m2	LD	0.45	9.9	0.006
battered backing	m2	LD	0.45	9.9	0.006

140mm thick

vertical straight walls	m2	LD	0.50	9.9	0.007
vertical curved walls	m2	LD	0.60	9.9	0.007
battered straight walls	m2	LD	0.60	9.9	0.007
battered curved walls	m2	LD	0.75	9.9	0.007
vertical backing	m2	LD	0.55	9.9	0.007
battered backing	m2	LD	0.55	9.9	0.007

215mm thick

vertical straight walls	m2	LD	0.60	9.9	0.007
vertical curved walls	m2	LD	0.70	9.9	0.007
battered straight walls	m2	LD	0.70	9.9	0.007
battered curved walls	m2	LD	0.85	9.9	0.007
vertical backing	m2	LD	0.65	9.9	0.007
battered backing	m2	LD	0.65	9.9	0.007

	Unit	Labour grade	Labour hours	Blocks nr	Mortar m3

**Dense blockwork
strength 7N/mm2, size
440 x 215mm in cement
mortar (1:3)**

100mm thick

	Unit	Labour grade	Labour hours	Blocks nr	Mortar m3
vertical straight walls	m2	LD	0.45	9.9	0.006
vertical curved walls	m2	LD	0.55	9.9	0.006
battered straight walls	m2	LD	0.55	9.9	0.006
battered curvedwalls	m2	LD	0.70	9.9	0.006
vertical backing	m2	LD	0.50	9.9	0.006
battered backing	m2	LD	0.50	9.9	0.006

140mm thick

	Unit	Labour grade	Labour hours	Blocks nr	Mortar m3
vertical straight walls	m2	LD	0.55	9.9	0.007
vertical curved walls	m2	LD	0.65	9.9	0.007
battered straight walls	m2	LD	0.65	9.9	0.007
battered curved walls	m2	LD	0.80	9.9	0.007
vertical backing	m2	LD	0.60	9.9	0.007
battered backing	m2	LD	0.60	9.9	0.007

215mm thick

	Unit	Labour grade	Labour hours	Blocks nr	Mortar m3
vertical straight walls	m2	LD	0.65	9.9	0.007
vertical curved walls	m2	LD	0.75	9.9	0.007
battered straight walls	m2	LD	0.75	9.9	0.007
battered curved walls	m2	LD	0.90	9.9	0.007
vertical backing	m2	LD	0.70	9.9	0.007
battered backing	m2	LD	0.70	9.9	0.007

	Unit	Labour grade	Labour hours	Stone tonne	Mortar m3
Masonry					
Ashlar masonry, natural Portland Whitbed, in cement mortar (1:3)					
75mm thick					
vertical straight walls	m2	LE	0.50	0.17	0.005
vertical curved walls	m2	LE	0.60	0.17	0.005
vertical backing	m2	LE	0.55	0.17	0.005
100mm thick					
vertical straight walls	m2	LE	0.60	0.22	0.006
vertical curved walls	m2	LE	0.70	0.22	0.006
vertical backing	m2	LE	0.65	0.22	0.006
150mm thick					
vertical straight walls	m2	LE	0.70	0.32	0.007
vertical curved walls	m2	LE	0.80	0.32	0.007
vertical backing	m2	LE	0.75	0.32	0.007
Cotswold stone random rubble walling, laid dry					
300mm thick					
vertical straight walls	m2	LE	0.80	0.60	-
vertical curved walls	m2	LE	0.90	0.60	-
vertical backing	m2	LE	0.85	0.60	-

	Unit	Labour grade	Labour hours	Stone tonne	Mortar m3
400mm thick					
vertical straight walls	m2	LE	1.00	0.80	-
vertical curved walls	m2	LE	1.10	0.80	-
vertical backing	m2	LE	1.05	0.80	-
450mm thick					
vertical straight walls	m2	LE	1.10	0.90	-
vertical curved walls	m2	LE	1.20	0.90	-
vertical backing	m2	LE	1.15	0.90	-
500mm thick					
vertical straight walls	m2	LE	1.20	1.00	-
vertical curved walls	m2	LE	1.30	1.00	-
vertical backing	m2	LE	1.25	1.00	-

Cotswold stone random rubble walling, laid in cement lime mortar (1:2:9)

	Unit	Labour grade	Labour hours	Stone tonne	Mortar m3
300mm thick					
vertical straight walls	m2	LE	0.90	0.60	0.12
vertical curved walls	m2	LE	1.00	0.60	0.12
vertical backing	m2	LE	0.95	0.60	0.12
400mm thick					
vertical straight walls	m2	LE	1.10	0.80	0.15
vertical curved walls	m2	LE	1.20	0.80	0.15
vertical backing	m2	LE	1.15	0.80	0.15

	Unit	Labour grade	Labour hours	Stone tonne	Mortar m3
450mm thick					
vertical straight walls	m2	LE	1.20	0.90	0.16
vertical curved walls	m2	LE	1.30	0.90	0.16
vertical backing	m2	LE	1.25	0.90	0.16
500mm thick					
vertical straight walls	m2	LE	1.30	1.00	0.17
vertical curved walls	m2	LE	1.40	1.00	0.17
vertical backing	m2	LE	1.35	1.00	0.17

Cotswold stone irregular coursed rubble walling, laid in cement lime mortar (1:2:9)

	Unit	Labour grade	Labour hours	Stone tonne	Mortar m3
300mm thick					
vertical straight walls	m2	LE	1.00	0.60	0.12
vertical curved walls	m2	LE	1.10	0.60	0.12
vertical backing	m2	LE	1.05	0.60	0.12
400mm thick					
vertical straight walls	m2	LE	1.20	0.80	0.15
vertical curved walls	m2	LE	1.30	0.80	0.15
vertical backing	m2	LE	1.25	0.80	0.15
450mm thick					
vertical straight walls	m2	LE	1.30	0.90	0.16
vertical curved walls	m2	LE	1.40	0.90	0.16
vertical backing	m2	LE	1.35	0.90	0.16

	Unit	Labour grade	Labour hours	Stone tonne	Mortar m3
500mm thick					
vertical straight walls	m2	LE	1.40	1.00	0.17
vertical curved walls	m2	LE	1.50	1.00	0.17
vertical backing	m2	LE	1.45	1.00	0.17

Cotswold stone coursed rubble walling, laid in cement lime mortar (1:2:9)

300mm thick					
vertical straight walls	m2	LE	0.90	0.60	0.12
vertical curved walls	m2	LE	1.00	0.60	0.12
vertical backing	m2	LE	0.95	0.60	0.12
400mm thick					
vertical straight walls	m2	LE	1.10	0.80	0.15
vertical curved walls	m2	LE	1.20	0.80	0.15
vertical backing	m2	LE	1.15	0.80	0.15
450mm thick					
vertical straight walls	m2	LE	1.20	0.90	0.16
vertical curved walls	m2	LE	1.30	0.90	0.16
vertical backing	m2	LE	1.25	0.90	0.16
500mm thick					
vertical straight walls	m2	LE	1.30	1.00	0.17
vertical curved walls	m2	LE	1.40	1.00	0.17
vertical backing	m2	LE	1.35	1.00	0.17

27

Timber

Weights of materials   **kg/m3**

Blockboard
standard	940-1000
tempered	940-1060

Wood chipboard
standard grade	650-750
flooring grade	680-800

Laminboard   500-700

Timber
Ash	800
Baltic Spruce	480
Beech	816
Birch	720
Box	961
Cedar	480
Ebony	1217
Elm	624
Greenheart	961
Jarrah	816
Maple	752
Oak, American	720
Oak, English	848
Pine, Pitchpine	800
Pine, Red Deal	576
Pine, Yellow Deal	528
Sycamore	530
Teak, African	961
Teak, Indian	656
Walnut	496

Number of nails per kg   nr

Oval brad or lost head nails

	nr
150 x 7.10 x 5.00	31
125 x 6.70 x 4.50	44
100 x 6.00 x 4.00	64
75 x 5.00 x 3.35	125
65 x 4.00 x 2.65	230
60 x 3.75 x 2.36	340
50 x 3.35 x 2.00	470
40 x 2.65 x 1.60	940
30 x 2.65 x 1.60	1480
25 x 2.00 x 1.25	2530

Round plain head nails

	nr
150 x 6.00	29
125 x 5.60	42
125 x 5.00	53
115 x 5.00	57
100 x 5.00	66
100 x 4.50	77
100 x 4.00	88
90 x 4.00	106
75 x 4.00	121
75 x 3.75	154
75 x 3.35	194
65 x 3.35	230
65 x 3.00	275
65 x 2.65	350
60 x 3.35	255
60 x 3.00	310
60 x 2.65	385
50 x 3.35	290
50 x 3.00	340
50 x 2.65	440
50 x 2.36	550
45 x 2.65	510
45 x 2.36	640
40 x 2.65	575

Number of nails per kg   nr

Round plain head nails

40 x 2.36	750
40 x 2.00	970
30 x 2.36	840
30 x 2.00	1170
25 x 2.00	1430
25 x 1.80	1720
25 x 1.60	2210
20 x 1.60	2710

Round lost head nails

65 x 3.35	240
65 x 3.00	270
75 x 3.75	160
60 x 3.35	270
60 x 3.00	330
50 x 3.00	360
40 x 2.36	760

Lengths of boarding required per m2

Effective width, mm	m/m2
75	13.33
100	10.00
125	8.00
150	6.67
175	5.71
200	5.00

Lengths of timber (metres per m3)

25 x 25mm	1600
25 x 50mm	800
25 x 75mm	533
25 x 100mm	400
25 x 125mm	320
25 x 150mm	267

Lengths of timber (metres per m3)

50 x 50mm	400
50 x 75mm	267
50 x 100mm	200
50 x 125mm	160
50 x 150mm	133
50 x 175mm	114
50 x 200mm	100
50 x 225mm	89
50 x 250mm	80
50 x 275mm	72
50 x 300mm	67
75 x 50mm	267
75 x 75mm	178
75 x 100mm	133
75 x 125mm	107
75 x 150mm	89
75 x 175mm	76
75 x 200mm	67
75 x 225mm	59
75 x 250mm	53
75 x 275mm	48
75 x 300mm	44
100 x 100mm	100
100 x 125mm	80
100 x 150mm	67
100 x 175mm	57
100 x 200mm	50
100 x 225mm	44
100 x 250mm	40
100 x 275mm	36
100 x 300mm	33
150 x 200mm	33
150 x 225mm	30
150 x 250mm	27
150 x 275mm	24
150 x 300mm	22

Lengths of timber (metres per m3)

200 x 200mm	25
200 x 225mm	22
200 x 250mm	20
200 x 275mm	18
200 x 300mm	17
250 x 250mm	16
250 x 275mm	15
250 x 300mm	13
300 x 300mm	11

Labour grades

2 Craftsman and 1 unskilled
operative   LI

Plant grades

1 Tractor trailer and
1 crawler crane   PZ

	Unit	Labour grade	Labour hours	Plant grade	Plant hours
Hardwood					
Greenheart, size 100 x 75mm, length					
not exceeding 1m	m	LI	0.26	PZ	0.26
1.5-3m	m	LI	0.24	PZ	0.24
3-5m	m	LI	0.22	PZ	0.22
Greenheart, size 150 x 75mm, length					
not exceeding 1m	m	LI	0.32	PZ	0.32
1.5-3m	m	LI	0.30	PZ	0.30
3-5m	m	LI	0.28	PZ	0.28
Greenheart, size 200 x 100mm, length					
not exceeding 1m	m	LI	0.48	PZ	0.48
1.5-3m	m	LI	0.45	PZ	0.45
3-5m	m	LI	0.42	PZ	0.42
Greenheart, size 200 x 150mm, length					
not exceeding 1m	m	LI	0.68	PZ	0.68
1.5-3m	m	LI	0.64	PZ	0.64
3-5m	m	LI	0.60	PZ	0.60
Greenheart, size 200 x 200mm, length					
not exceeding 1m	m	LI	0.92	PZ	0.92
1.5-3m	m	LI	0.86	PZ	0.86
3-5m	m	LI	0.80	PZ	0.80

	Unit	Labour grade	Labour hours	Plant grade	Plant hours
Greenheart, size 200 x 300mm, length					
not exceeding 1m	m	LI	1.20	PZ	1.20
1.5-3m	m	LI	1.12	PZ	1.12
3-5m	m	LI	1.04	PZ	1.04
Greenheart, size 300 x 300mm, length					
not exceeding 1m	m	LI	1.54	PZ	1.54
1.5-3m	m	LI	1.44	PZ	1.44
3-5m	m	LI	1.34	PZ	1.34
Greenheart, size 600 x 300mm, length					
not exceeding 1m	m	LI	1.74	PZ	1.74
1.5-3m	m	LI	1.64	PZ	1.64
3-5m	m	LI	1.54	PZ	1.54
Greenheart, size 600 x 450mm, length					
not exceeding 1m	m	LI	2.10	PZ	2.10
1.5-3m	m	LI	1.95	PZ	1.95
3-5m	m	LI	1.80	PZ	1.80
Greenheart, size 600 x 600mm, length					
not exceeding 1m	m	LI	2.50	PZ	2.50
1.5-3m	m	LI	2.30	PZ	2.30
3-5m	m	LI	2.10	PZ	2.10

	Unit	Labour grade	Labour hours	Plant grade	Plant hours

Softwood

Douglas fir, size 50 x 75mm, length

not exceeding 1m	m	LI	0.12	PZ	0.12
1.5-3m	m	LI	0.10	PZ	0.10
3-5m	m	LI	0.08	PZ	0.08

Douglas fir, size 150 x 75mm, length

not exceeding 1m	m	LI	0.22	PZ	0.22
1.5-3m	m	LI	0.18	PZ	0.18
3-5m	m	LI	0.14	PZ	0.14

Douglas fir, size 200 x 150mm, length

not exceeding 1m	m	LI	0.52	PZ	0.52
1.5-3m	m	LI	0.46	PZ	0.46
3-5m	m	LI	0.40	PZ	0.40

Douglas fir, size 200 x 200mm, length

not exceeding 1m	m	LI	0.68	PZ	0.68
1.5-3m	m	LI	0.60	PZ	0.60
3-5m	m	LI	0.52	PZ	0.52

Douglas fir, size 300 x 300mm, length

not exceeding 1m	m	LI	0.92	PZ	0.92
1.5-3m	m	LI	0.82	PZ	0.82
3-5m	m	LI	0.72	PZ	0.72

	Unit	Labour grade	Labour hours	Plant grade	Plant hours
Douglas fir, size 600 x 300mm, length					
not exceeding 1m	m	LI	1.30	PZ	1.30
1.5-3m	m	LI	1.20	PZ	1.20
3-5m	m	LI	1.10	PZ	1.10
Douglas fir, size 600 x 450mm, length					
not exceeding 1m	m	LI	1.50	PZ	1.50
1.5-3m	m	LI	1.35	PZ	1.35
3-5m	m	LI	1.20	PZ	1.20
Douglas fir, size 600 x 600mm, length					
not exceeding 1m	m	LI	2.10	PZ	2.10
1.5-3m	m	LI	1.90	PZ	1.90
3-5m	m	LI	1.70	PZ	1.70

Decking

	Unit	Labour grade	Labour hours	Plant grade	Plant hours
Greenheart decking, thickness					
25mm	m2	LI	0.50	PZ	0.50
50mm	m2	LI	0.70	PZ	0.70
75mm	m2	LI	0.90	PZ	0.90
100mm	m2	LI	1.10	PZ	1.10
Douglas fir decking, thickness					
25mm	m2	LI	0.40	PZ	0.40
50mm	m2	LI	0.60	PZ	0.60
75mm	m2	LI	0.80	PZ	0.80
100mm	m2	LI	0.90	PZ	0.90

	Unit	Labour grade	Labour hours	Plant grade	Plant hours

Fittings and fastenings

Galvanised mild steel
straps, size

	Unit	Labour grade	Labour hours	Plant grade	Plant hours
30 x 2.5 x 300mm girth	nr	PI	0.16	-	-
30 x 2.5 x 400mm girth	nr	PI	0.17	-	-
30 x 2.5 x 500mm girth	nr	PI	0.18	-	-
30 x 2.5 x 600mm girth	nr	PI	0.19	-	-
30 x 2.5 x 800mm girth	nr	PI	0.20	-	-
30 x 2.5 x 900mm girth	nr	PI	0.21	-	-
30 x 2.5 x 1000mm girth	nr	PI	0.22	-	-
30 x 2.5 x 1100mm girth	nr	PI	0.23	-	-
30 x 2.5 x 1200mm girth	nr	PI	0.24	-	-

Galvanised mild steel
rosehead spikes, size

	Unit	Labour grade	Labour hours	Plant grade	Plant hours
12 x 12 x 100mm long	nr	PI	0.16	-	-
12 x 12 x 250mm long	nr	PI	0.17	-	-
12 x 15 x 250mm long	nr	PI	0.18	-	-

Steel coach screws with
square head, 6mm
diameter, length

	Unit	Labour grade	Labour hours	Plant grade	Plant hours
50mm	nr	PI	0.02	-	-
75mm	nr	PI	0.02	-	-
100mm	nr	PI	0.03	-	-
125mm	nr	PI	0.03	-	-

Steel coach screws with
square head, 8mm
diameter, length

	Unit	Labour grade	Labour hours	Plant grade	Plant hours
50mm	nr	PI	0.03	-	-
75mm	nr	PI	0.03	-	-
100mm	nr	PI	0.04	-	-
125mm	nr	PI	0.04	-	-

	Unit	Labour grade	Labour hours	Plant grade	Plant hours
Steel coach screws with square head, 10mm diameter, length					
50mm	nr	PI	0.04	-	-
75mm	nr	PI	0.04	-	-
100mm	nr	PI	0.05	-	-
125mm	nr	PI	0.05	-	-
150mm	nr	PI	0.05	-	-
Steel coach screws with square head, 10mm diameter, length					
100mm	nr	PI	0.06	-	-
125mm	nr	PI	0.06	-	-
150mm	nr	PI	0.06	-	-
Steel coach screws with square head, 12.5mm diameter, length					
100mm	nr	PI	0.07	-	-
125mm	nr	PI	0.07	-	-
150mm	nr	PI	0.07	-	-
M6 black bolts with hexagonal heads each with nut and washer					
25mm long	nr	PI	0.05	-	-
50mm long	nr	PI	0.05	-	-
75mm long	nr	PI	0.05	-	-
100mm long	nr	PI	0.05	-	-

	Unit	Labour grade	Labour hours	Plant grade	Plant hours
M8 black bolts with hexagonal heads each with nut and washer					
25mm long	nr	PI	0.05	-	-
50mm long	nr	PI	0.05	-	-
75mm long	nr	PI	0.05	-	-
100mm long	nr	PI	0.05	-	-
M10 black bolts with hexagonal heads each with nut and washer					
50mm long	nr	PI	0.06	-	-
75mm long	nr	PI	0.06	-	-
100mm long	nr	PI	0.08	-	-
125mm long	nr	PI	0.08	-	-
150mm long	nr	PI	0.08	-	-
M12 black bolts with hexagonal heads each with nut and washer					
50mm long	nr	PI	0.06	-	-
75mm long	nr	PI	0.06	-	-
100mm long	nr	PI	0.08	-	-
125mm long	nr	PI	0.08	-	-
150mm long	nr	PI	0.10	-	-
200mm long	nr	PI	0.10	-	-
250mm long	nr	PI	0.12	-	-
300mm long	nr	PI	0.12	-	-
M16 black bolts with hexagonal heads each with nut and washer					
50mm long	nr	PI	0.06	-	-
75mm long	nr	PI	0.06	-	-
100mm long	nr	PI	0.08	-	-

	Unit	Labour grade	Labour hours	Plant grade	Plant hours
Galvanised steel single-sided timber connectors with round toothed plate for 10mm diameter bolts, diameter					
38mm	nr	PI	0.02	-	-
50mm	nr	PI	0.02	-	-
63mm	nr	PI	0.02	-	-
75mm	nr	PI	0.02	-	-
Galvanised steel single-sided timber connectors with round toothed plate for 12mm diameter bolts, diameter					
38mm	nr	PI	0.02	-	-
50mm	nr	PI	0.02	-	-
63mm	nr	PI	0.02	-	-
75mm	nr	PI	0.02	-	-
Galvanised steel double-sided timber connectors with round toothed plate for 10mm diameter bolts, diameter					
38mm	nr	PI	0.03	-	-
50mm	nr	PI	0.03	-	-
63mm	nr	PI	0.03	-	-
75mm	nr	PI	0.03	-	-

	Unit	Labour grade	Labour hours	Plant grade	Plant hours
Galvanised steel double-sided timber connectors with round toothed plate for 12mm diameter bolts, diameter					
38mm	nr	PI	0.03	-	-
50mm	nr	PI	0.03	-	-
63mm	nr	PI	0.03	-	-
75mm	nr	PI	0.03	-	-
Galvanised mild steel split ring connectors, diameter					
50mm	nr	PI	0.10	-	-
75mm	nr	PI	0.12	-	-

28

Painting

Average coverage of paints in square metres per litre

Surfaces	A	B	C	D	E	F	G	H
Lead-based primer	13-15	-	-	-	-	-	10-12	7-10
Zinc-based primer	13-15	-	-	-	-	-	10-12	7-10
Etch-based primer	13-15	-	-	-	-	-	10-12	7-10
Oil paint	11-14	8-10	8-10	7-9	6-8	-	11-14	10-12
Emulsion paint	12-15	8-12	11-14	8-12	6-10	2-4	12-15	8-10
Cement paint	-	4-6	6-7	3-6	3-6	2-3	-	-

Surfaces	I	J	K	L	M	N	O	P
Lead-based primer	-	8-11	7-10	-	-	10-14	-	-
Zinc-based primer	-	8-11	7-10	-	-	10-14	-	-
Etch-based primer	-	8-11	7-10	-	-	10-14	-	-
Oil paint	10-12	11-14	10-12	10-12	10-12	10-12	11-14	11-14
Gloss finish	10-12	11-14	10-12	10-12	10-12	10-12	11-14	11-14
Emulsion paint	8-10	12-15	10-12	-	-	10-12	12-15	12-15
Cement paint	-	-	4-6	-	-	-	-	-

(See key on following page)

Key

A - Finishing plaster
B - Wood-floated rendering
C - Smooth concrete/cement
D - Fair-faced brickwork
E - Blockvork
F - Roughcast/pebble dash
G - Hardboard
H - Soft fibre insulating board
I - Fire-retardant fibre insulating board
J - Smooth paper-faced board
K - Hard asbestos sheet
L - Structural steelwork
M - Metal sheeting
N - Joinery
O - Smooth primed surfaces
P - Smooth undercoated surfaces

Labour grades

1 Craftsman   LA

	Unit	Labour grade	Labour hours	Paint litres
Primers				
One coat lead-based primer on metal surfaces				
exceeding 1m	m2	LA	0.24	0.10
not exceeding 0.3m	m	LA	0.08	0.03
0.3-1m	m	LA	0.15	0.05
One coat lead-based primer on metal pipework, nominal bore				
not exceeding 50mm	m	LA	0.05	0.01
50-100mm	m	LA	0.10	0.03
100-200mm	m	LA	0.20	0.06
300mm	m	LA	0.30	0.10
One coat zinc-based primer on metal surfaces				
exceeding 1m	m2	LA	0.22	0.10
not exceeding 0.3m	m	LA	0.07	0.03
0.3-1m	m	LA	0.12	0.05
One coat zinc-based primer on metal pipework, nominal bore				
not exceeding 50mm	m	LA	0.04	0.01
50-100mm	m	LA	0.08	0.03
100-200mm	m	LA	0.18	0.06
300mm	m	LA	0.26	0.10

	Unit	Labour grade	Labour hours	Paint litres
One coat etch-based primer on metal surfaces				
exceeding 1m	m2	LA	0.24	0.10
not exceeding 0.3m	m	LA	0.08	0.03
0.3-1m	m	LA	0.14	0.05
One coat zinc-based primer on metal pipework, nominal bore				
not exceeding 50mm	m	LA	0.05	0.01
50-100mm	m	LA	0.10	0.03
100-200mm	m	LA	0.20	0.06
300mm	m	LA	0.30	0.10

Oil paint

	Unit	Labour grade	Labour hours	Paint litres
One coat oil paint on primed metal surfaces				
exceeding 1m	m2	LA	0.24	0.09
not exceeding 0.3m	m	LA	0.08	0.03
0.3-1m	m	LA	0.14	0.05
One coat oil paint on primed wood surfaces				
exceeding 1m	m2	LA	0.24	0.09
not exceeding 0.3m	m	LA	0.08	0.03
0.3-1m	m	LA	0.14	0.05

	Unit	Labour grade	Labour hours	Paint litres
One coat oil paint on metal pipework, nominal bore				
not exceeding 50mm	m	LA	0.05	0.01
50-100mm	m	LA	0.10	0.03
100-200mm	m	LA	0.20	0.06
300mm	m	LA	0.30	0.10
Two coats oil paint on primed metal surfaces				
exceeding 1m	m2	LA	0.40	0.18
not exceeding 0.3m	m	LA	0.14	0.06
0.3-1m	m	LA	0.20	0.10
Two coats oil paint on primed wood surfaces				
exceeding 1m	m2	LA	0.40	0.18
not exceeding 0.3m	m	LA	0.14	0.06
0.3-1m	m	LA	0.20	0.10
Two coats oil paint on metal pipework, nominal bore				
not exceeding 50mm	m	LA	0.08	0.02
50-100mm	m	LA	0.18	0.06
100-200mm	m	LA	0.34	0.12
300mm	m	LA	0.50	0.20
Three coats oil paint on primed metal surfaces				
exceeding 1m	m2	LA	0.56	0.27
not exceeding 0.3m	m	LA	0.20	0.09
0.3-1m	m	LA	0.30	0.15

	Unit	Labour grade	Labour hours	Paint litres
Three coats oil paint on primed wood surfaces				
exceeding 1m	m2	LA	0.56	0.27
not exceeding 0.3m	m	LA	0.20	0.09
0.3-1m	m	LA	0.30	0.15
Three coats oil paint on metal pipework, nominal bore				
not exceeding 50mm	m	LA	0.12	0.03
50-100mm	m	LA	0.24	0.09
100-200mm	m	LA	0.42	0.18
300mm	m	LA	0.60	0.30

Emulsion paint

	Unit	Labour grade	Labour hours	Paint litres
One coat emulsion paint on primed concrete surfaces				
exceeding 1m	m2	LA	0.16	0.04
not exceeding 0.3m	m	LA	0.05	0.01
0.3-1m	m	LA	0.08	0.02
One coat emulsion paint on primed masonry surfaces				
exceeding 1m	m2	LA	0.18	0.04
not exceeding 0.3m	m	LA	0.06	0.01
0.3-1m	m	LA	0.09	0.02

	Unit	Labour grade	Labour hours	Paint litres
One coat emulsion paint on primed brick surfaces				
exceeding 1m	m2	LA	0.16	0.04
not exceeding 0.3m	m	LA	0.05	0.01
0.3-1m	m	LA	0.08	0.02
Two coats emulsion paint on primed concrete surfaces				
exceeding 1m	m2	LA	0.26	0.08
not exceeding 0.3m	m	LA	0.08	0.02
0.3-1m	m	LA	0.12	0.04
Two coats emulsion paint on primed masonry surfaces				
exceeding 1m	m2	LA	0.30	0.08
not exceeding 0.3m	m	LA	0.10	0.02
0.3-1m	m	LA	0.14	0.04
Two coats emulsion paint on primed brick surfaces				
exceeding 1m	m2	LA	0.26	0.08
not exceeding 0.3m	m	LA	0.08	0.02
0.3-1m	m	LA	0.12	0.04
Three coats emulsion paint on primed concrete surfaces				
exceeding 1m	m2	LA	0.36	0.12
not exceeding 0.3m	m	LA	0.10	0.03
0.3-1m	m	LA	0.16	0.05

	Unit	Labour grade	Labour hours	Paint litres
Three coats emulsion paint on primed masonry surfaces				
exceeding 1m	m2	LA	0.42	0.12
not exceeding 0.3m	m	LA	0.14	0.03
0.3-1m	m	LA	0.20	0.05
Three coats emulsion paint on primed brick surfaces				
exceeding 1m	m2	LA	0.26	0.12
not exceeding 0.3m	m	LA	0.08	0.03
0.3-1m	m	LA	0.12	0.05

Masonry paint

	Unit	Labour grade	Labour hours	Paint litres
One coat cement paint on primed concrete surfaces				
exceeding 1m	m2	LA	0.16	0.03
not exceeding 0.3m	m	LA	0.05	0.01
0.3-1m	m	LA	0.08	0.02
One coat cement paint on primed masonry surfaces				
exceeding 1m	m2	LA	0.18	0.03
not exceeding 0.3m	m	LA	0.06	0.01
0.3-1m	m	LA	0.09	0.02
One coat cement paint on primed brick surfaces				
exceeding 1m	m2	LA	0.16	0.03
not exceeding 0.3m	m	LA	0.05	0.01
0.3-1m	m	LA	0.08	0.02

	Unit	Labour grade	Labour hours	Paint litres
Two coats cement paint on primed concrete surfaces				
exceeding 1m	m2	LA	0.26	0.08
not exceeding 0.3m	m	LA	0.08	0.02
0.3-1m	m	LA	0.12	0.04
Two coats cement paint on primed masonry surfaces				
exceeding 1m	m2	LA	0.30	0.08
not exceeding 0.3m	m	LA	0.10	0.02
0.3-1m	m	LA	0.14	0.04
Two coats cement paint on primed brick surfaces				
exceeding 1m	m2	LA	0.26	0.08
not exceeding 0.3m	m	LA	0.08	0.02
0.3-1m	m	LA	0.12	0.04
Three coats cement paint on primed concrete surfaces				
exceeding 1m	m2	LA	0.36	0.12
not exceeding 0.3m	m	LA	0.10	0.03
0.3-1m	m	LA	0.16	0.05
Three coats cement paint on primed masonry surfaces				
exceeding 1m	m2	LA	0.42	0.12
not exceeding 0.3m	m	LA	0.14	0.03
0.3-1m	m	LA	0.20	0.05

	Unit	Labour grade	Labour hours	Paint litres
Three coats emulsion paint on primed brick surfaces				
exceeding 1m	m2	LA	0.26	0.12
not exceeding 0.3m	m	LA	0.08	0.03
0.3-1m	m	LA	0.12	0.05

Varnish

	Unit	Labour grade	Labour hours	Paint litres
One coat polyurethane on primed wood surfaces				
exceeding 1m	m2	LA	0.22	0.03
not exceeding 0.3m	m	LA	0.06	0.01
0.3-1m	m	LA	0.12	0.02
Two coats polyurethane on primed wood surfaces				
exceeding 1m	m2	LA	0.36	0.06
not exceeding 0.3m	m	LA	0.10	0.02
0.3-1m	m	LA	0.20	0.04
Three coats polyurethane on primed wood surfaces				
exceeding 1m	m2	LA	0.50	0.09
not exceeding 0.3m	m	LA	0.14	0.03
0.3-1m	m	LA	0.26	0.06

Bituminous paint

	Unit	Labour grade	Labour hours	Paint litres
One coat bituminous paint on primed wood surfaces				
exceeding 1m	m2	LA	0.24	0.09
not exceeding 0.3m	m	LA	0.08	0.03
0.3-1m	m	LA	0.13	0.05

	Unit	Labour grade	Labour hours	Paint litres
One coat bituminous paint on primed concrete surfaces				
exceeding 1m	m2	LA	0.26	0.09
not exceeding 0.3m	m	LA	0.09	0.03
0.3-1m	m	LA	0.13	0.05
One coat bituminous paint on primed metal surfaces				
exceeding 1m	m2	LA	0.24	0.09
not exceeding 0.3m	m	LA	0.08	0.03
0.3-1m	m	LA	0.13	0.05
Two coats bituminous paint on primed wood surfaces				
exceeding 1m	m2	LA	0.40	0.18
not exceeding 0.3m	m	LA	0.14	0.06
0.3-1m	m	LA	0.20	0.10
Two coats bituminous paint on primed concrete surfaces				
exceeding 1m	m2	LA	0.44	0.18
not exceeding 0.3m	m	LA	0.16	0.06
0.3-1m	m	LA	0.22	0.10
Two coats bituminous paint on primed metal surfaces				
exceeding 1m	m2	LA	0.40	0.18
not exceeding 0.3m	m	LA	0.14	0.06
0.3-1m	m	LA	0.20	0.10

29

Waterproofing

Weights of materials

Lead sheeting, 2.65mm thick	30.10kg/m2
Lead sheeting, 3.55mm thick	40.26kg/m2
Zinc sheeting, 0.8mm thick	4.00kg/m2
Asphalt, 20mm thick	0.05tonnes/m2
Cement and sand render 20mm thick	
Cement	40.00kg/m2
Sand	9.00kg/m2
Liquid waterproofing solution	0.85litres/m2

Labour grades

1 Craftsman	LA
2 Asphalt layers and 1 unskilled operative	LF

Plant grades

Asphalt boiler	PG

	Unit	Labour grade	Labour hours	Plant grade	Plant hours	Materials tonne

Damp proofing

Two coat asphalt work
20mm thick, width

	Unit	Labour grade	Labour hours	Plant grade	Plant hours	Materials tonne
exceeding 1m	m2	LF	0.40	PG	0.40	0.045
not exceeding 0.3m	m2	LF	0.15	PG	0.15	0.015
0.3-1m	m2	LF	0.20	PG	0.20	0.022

Sheet lead code 6,
colour-coded black

	Unit	Labour grade	Labour hours	Plant grade	Plant hours	Materials tonne
exceeding 1m	m2	LF	4.40	-	-	0.031
not exceeding 0.3m	m2	LF	0.55	-	-	0.010
0.3-1m	m2	LF	1.20	-	-	0.015

	Unit	Labour grade	Labour hours	Mortar m3

Two coat cement and
sand (1:3) mixed with a
waterproofing additive,
20mm thick, laid
horizontally, width

	Unit	Labour grade	Labour hours	Mortar m3
exceeding 1m	m2	LA	0.70	0.019
not exceeding 0.3m	m2	LA	0.22	0.006
0.3-1m	m2	LA	0.35	0.009

	Unit	Labour grade	Labour hours	Plant grade	Plant hours	Materials tonne

Tanking

Three coat asphalt work
26mm thick, width

	Unit	Labour grade	Labour hours	Plant grade	Plant hours	Materials tonne
exceeding 1m	m2	LF	0.50	PG	0.50	0.060
not exceeding 0.3m	m	LF	0.18	PG	0.18	0.020
0.3-1m	m	LF	0.25	PG	0.25	0.030

	Unit	Labour grade	Labour hours	Mortar m3

Two coat cement and
sand (1:3) mixed with a
waterproofing additive,
20mm thick, laid vertically,
width

	Unit	Labour grade	Labour hours	Mortar m3
exceeding 1m	m2	LA	0.70	0.019
not exceeding 0.3m	m	LA	0.22	0.006
0.3-1m	m	LA	0.35	0.009

Protective layers

Flexible polythene sheeting,
1200 grade, width

	Unit	Labour grade	Labour hours	Mortar m3
exceeding 1m	m2	LA	0.03	-
not exceeding 0.3m	m	LA	0.01	-
0.3-1m	m	LA	0.02	-

Flexible polythene sheeting,
4000 grade, width

	Unit	Labour grade	Labour hours	Mortar m3
exceeding 1m	m2	LA	0.05	-
not exceeding 0.3m	m	LA	0.02	-
0.3-1m	m	LA	0.03	-

30

Fencing

Labour grades

Semi-skilled operative   LB

	Unit	Labour grade	Labour hours	Rails m	Posts nr	Concrete m3

Timber fencing

Chestnut fencing with
pales at 75mm centres
on two lines of galvanised
steel wire fixed to 75mm
diameter at 3m centres,
driven into ground,
height

900mm	m	LA	0.45	-	0.33	-
1000mm	m	LA	0.48	-	0.33	-
1100mm	m	LA	0.50	-	0.33	-
1200mm	m	LA	0.52	-	0.33	-

Chestnut fencing with
pales at 75mm centres
on two lines of galvanised
steel wire fixed to 75mm
diameter at 3m centres,
driven into ground,
height

900mm	m	LA	0.48	-	0.33	-
1000mm	m	LA	0.50	-	0.33	-
1100mm	m	LA	0.52	-	0.33	-
1200mm	m	LA	0.54	-	0.33	-

Timber post and rail
fencing consisting of
100mm posts at 2m
centres set in concrete,
two 75mm half round
rails, height

1000mm	m	LA	0.60	2.00	0.50	0.02
1100mm	m	LA	0.65	2.00	0.50	0.02

	Unit	Labour grade	Labour hours	Rails m	Posts nr	Concrete m3
Timber post and rail fencing consisting of 100mm posts at 2m centres set in concrete, three 75mm half round rails, height						
1100mm	m	LA	0.75	3.00	0.50	0.02
1200mm	m	LA	0.80	3.00	0.50	0.02
Timber post and rail fencing consisting of 100mm posts at 2m centres set in concrete, four 75mm half round rails, height						
1200mm	m	LA	0.90	4.00	0.50	0.02
1300mm	m	LA	0.95	4.00	0.50	0.02

	Unit	Labour grade	Labour hours	Wire m	Posts nr	Concrete m3
Timber post and wire fencing consisting of 100mm posts at 2m centres set in concrete, rails, height 1200mm						
three strands of galvanised wire	m	LA	0.35	3.00	0.50	0.02
three strands of single barbed wire	m	LA	0.40	3.00	0.50	0.02
three strands of double barbed wire	m	LA	0.45	3.00	0.50	0.02

	Unit	Labour grade	Labour hours	Wire m	Posts nr	Concrete m3
Timber post and wire fencing consisting of 100mm posts at 2m centres set in concrete, rails, height 1350mm						
three strands of galvanised wire	m	LA	0.45	3.00	0.50	0.02
three strands of single barbed wire	m	LA	0.50	3.00	0.50	0.02
three strands of double barbed wire	m	LA	0.55	3.00	0.50	0.02
Timber post and wire fencing consisting of 100mm posts at 2m centres set in concrete, rails, height 1500mm						
three strands of galvanised wire	m	LA	0.55	3.00	0.50	0.02
three strands of single barbed wire	m	LA	0.60	3.00	0.50	0.02
three strands of double barbed wire	m	LA	0.65	3.00	0.50	0.02
Timber post and wire fencing consisting of 100mm posts at 2m centres set in concrete, rails, height 1750mm						
three strands of galvanised wire	m	LA	0.65	3.00	0.50	0.02
three strands of single barbed wire	m	LA	0.60	3.00	0.50	0.02
three strands of double barbed wire	m	LA	0.65	3.00	0.50	0.02

	Unit	Labour grade	Labour hours	Mesh m2	Posts nr	Concrete m3

Metal fencing

Chainlink fencing with
2.5mm galvanised mesh
wire, line and tying wire,
fixed to concrete posts set
in concrete at 3m centres,
height

	Unit	Labour grade	Labour hours	Mesh m2	Posts nr	Concrete m3
900mm	m	LB	0.50	0.90	0.33	0.02
1200mm	m	LB	0.55	1.20	0.33	0.02
1500mm	m	LB	0.60	1.40	0.33	0.02
1800mm	m	LB	0.70	1.80	0.33	0.02
2100mm	m	LB	0.80	2.10	0.33	0.02
2400mm	m	LB	0.90	2.40	0.33	0.02

Chainlink fencing with
2.5mm plastic coated mesh
wire, line and tying wire,
fixed to concrete posts set
in concrete at 3m centres,
height

	Unit	Labour grade	Labour hours	Mesh m2	Posts nr	Concrete m3
900mm	m	LB	0.50	0.90	0.33	0.02
1200mm	m	LB	0.55	1.20	0.33	0.02
1500mm	m	LB	0.60	1.40	0.33	0.02
1800mm	m	LB	0.70	1.80	0.33	0.02
2100mm	m	LB	0.80	2.10	0.33	0.02
2400mm	m	LB	0.90	2.40	0.33	0.02

	Unit	Labour grade	Labour hours	Mesh m2	Posts nr	Concrete m3

Chainlink fencing with
2.5mm galvanised
mesh wire, line and
tying wire, fixed to
concrete posts set in
concrete at 3m centres,
height

	Unit	Labour grade	Labour hours	Mesh m2	Posts nr	Concrete m3
900mm	m	LB	0.45	0.90	0.33	0.02
1200mm	m	LB	0.50	1.20	0.33	0.02
1500mm	m	LB	0.55	1.40	0.33	0.02
1800mm	m	LB	0.65	1.80	0.33	0.02
2100mm	m	LB	0.75	2.10	0.33	0.02
2400mm	m	LB	0.85	2.40	0.33	0.02

Chainlink fencing with
2.5mm plastic coated
mesh wire, line and
tying wire, fixed to
concrete posts set in
concrete at 3m centres,
height

	Unit	Labour grade	Labour hours	Mesh m2	Posts nr	Concrete m3
900mm	m	LB	0.45	0.90	0.33	0.02
1200mm	m	LB	0.50	1.20	0.33	0.02
1500mm	m	LB	0.55	1.40	0.33	0.02
1800mm	m	LB	0.65	1.80	0.33	0.02
2100mm	m	LB	0.75	2.10	0.33	0.02
2400mm	m	LB	0.85	2.40	0.33	0.02

	Unit	Labour grade	Labour hours	Wire m	Posts nr	Concrete m3
Galvanised metal post and wire fencing consisting of 50 x 50mm angle iron posts at 3m centres holed for line wires, height 1350mm						
three strands of galvanised wire	m	LA	0.35	9.00	0.33	0.02
three strands of single barbed wire	m	LA	0.40	9.00	0.33	0.02
three strands of double barbed wire	m	LA	0.45	9.00	0.33	0.02
Galvanised metal post and wire fencing consisting of 50 x 50mm angle iron posts at 3m centres holed for line wires, height 1500mm						
four strands of galvanised wire	m	LA	0.40	12.00	0.33	0.02
four strands of single barbed wire	m	LA	0.45	12 .00	0.33	0.02
four strands of double barbed wire	m	LA	0.50	12.00	0.33	0.02
Galvanised metal post and wire fencing consisting of 50 x 50mm angle iron posts at 3m centres holed for line wires, height 1500mm						
five strands of galvanised wire	m	LA	0.45	15.00	0.33	0.02
five strands of single barbed wire	m	LA	0.50	15.00	0.33	0.02
five strands of double barbed wire	m	LA	0.55	15.00	0.33	0.02

	Unit	Labour grade	Labour hours	Mesh m2	Posts nr	Concrete m3
Galvanised metal post and wire fencing consisting of 50 x 50mm angle iron posts at 3m centres holed for line wires, height 1800mm						
five strands of galvanised wire	m	LA	0.50	15.00	0.50	0.02
five strands of single barbed wire	m	LA	0.55	15.00	0.50	0.02
five strands of double barbed wire	m	LA	0.60	15.00	0.50	0.02
Galvanised metal post and wire fencing consisting of 50 x 50mm angle iron posts at 3m centres holed for line wires, height 2100mm						
six strands of galvanised wire	m	LA	0.60	18.00	0.50	0.02
six strands of single barbed wire	m	LA	0.65	18.00	0.50	0.02
six strands of double barbed wire	m	LA	0.70	18.00	0.50	0.02

	Unit	Labour grade	Labour hours	Posts nr	Concrete m3

Gates

	Unit	Labour grade	Labour hours	Posts nr	Concrete m3
Treated softwood 5 bar framed and braced field gate hung on two 200 x 200mm posts surrounded in concrete, size					
2400 x 1125mm	nr	LA	7.00	2.00	0.05
3000 x 1125mm	nr	LA	7.50	2.00	0.05
3600 x 1125mm	nr	LA	8.00	2.00	0.05

	Unit	Labour grade	Labour hours	Posts nr	Concrete m3
Metal gates for use in chainlink fencing consisting of galvanised steel circular frame with galvanised steel infill panel, complete with fittings fixed to concrete posts surrounded in concrete, size					
900 x 900mm	nr	LA	1.00	2.00	0.05
900 x 1200mm	nr	LA	1.10	2.00	0.05
900 x 1400mm	nr	LA	1.20	2.00	0.05
900 x 1800mm	nr	LA	1.30	2.00	0.05
900 x 2100mm	nr	LA	1.40	2.00	0.05
900 x 2400mm	nr	LA	1.50	2.00	0.05

31

Gabions

Weights of materials

Rock   1750kg/m3

Labour grades

Semi-skilled operative   LB

Plant grades

Hydraulic excavator (1.7m3)   PA

	Unit	Labour grade	Labour hours	Plant grade	Plant hours	Rock tonnes
Galvanised wire mesh box gabions filled with rock and placed in position, size						
1.5 x 1 x 1m	nr	LB	1.20	PA	1.20	2.625
2 x 1 x 1m	nr	LB	1.60	PA	1.60	3.500
3 x 1 x 1m	nr	LB	2.40	PA	2.40	5.250
2 x 1 x 0.5m	nr	LB	0.80	PA	0.80	0.875
Galvanised wire mesh box mattresses filled with rock and placed in position, size						
6 x 2 x 0.17m	nr	LB	1.60	PA	1.60	3.570
6 x 2 x 0.23m	nr	LB	2.20	PA	2.20	3.865
6 x 2 x 0.3m	nr	LB	2.90	PA	2.90	5.050

PART THREE

LANDSCAPING

32

Seeding, soiling and turfing

Average weights of materials	kg/m3
Clay dry	1800
Topsoil	1600
Water	950

Quantities of seed required for sportsfields

Sport	Size m	Area m2	grams/m2 34kg	50kg	102kg	500kg	Weedkiller litres/ha
Bowling green	38.4x38.4	1475	50	75	150	200	1.475
Cricket square	22.8x22.8	522	18	27	54	72	0.522
Golf green (each)	-	570	20	30	58	78	0.570
Tennis	36.6x18.3	670	23	35	69	92	0.670
Football	119x91	10380	368	552	1104	1472	10.830
Rugby	100x69	6900	235	352	705	940	6.900
Hockey	91x55	5005	170	255	510	680	5.000
Hectare	-	10000	340	510	1020	1360	10.000
Acre	-	4047	137	206	413	550	24.050

Labour grades

Semi-skilled operator	**LB**
Unskilled operative	**LC**

Plant grades

Tractor and harrow	**PH**
Tractor and seeder	**PI**
Hand roller	**PJ**

	Unit	Labour grade	Labour hours	Plant grade	Plant hours	Materials tonnes

Pre-seeding work (by hand)

Lift topsoil from spoil
heap and spread in
layers, thickness

50mm	m2	LC	0.033	-	-	0.08
75mm	m2	LC	0.045	-	-	0.12
100mm	m2	LC	0.056	-	-	0.16

Rake topsoil to fine
tilth, depth

50mm	m2	LC	0.007	-	-	-
75mm	m2	LC	0.008	-	-	-
100mm	m2	LC	0.009	-	-	-

Pre-seeding work (by machine)

Lift topsoil from spoil
heap and spread in
layers, thickness

50mm	m2	-	-	PH	0.01	0.08
75mm	m2	-	-	PH	0.02	0.12
100mm	m2	-	-	PH	0.03	0.16

Harrow topsoil to
fine tilth, depth

50mm	m2	-	-	PH	0.01	-
75mm	m2	-	-	PH	0.02	-
100mm	m2	-	-	PH	0.03	-

	Unit	Labour grade	Labour hours	Plant grade	Plant hours	Materials kg
Harrow topsoil to fine tilth, depth						
50mm	m2	-	-	PH	0.01	-
75mm	m2	-	-	PH	0.02	-
100mm	m2	-	-	PH	0.03	-

Grass seeding (by hand)

Sow grass seed in two operations, grammes per m2

	Unit	Labour grade	Labour hours	Plant grade	Plant hours	Materials kg
12	100 m2	LC	1.00	-	-	1.20
15	100 m2	LC	1.00	-	-	1.50
20	100 m2	LC	1.00	-	-	2.00
25	100 m2	LC	1.00	-	-	2.50
30	100 m2	LC	1.00	-	-	3.00
35	100 m2	LC	1.00	-	-	3.50
40	100 m2	LC	1.00	-	-	4.00
50	100 m2	LC	1.00	-	-	5.00

Grass seeding (by machine)

Sow grass seed in two operations, grammes per m2

	Unit	Labour grade	Labour hours	Plant grade	Plant hours	Materials kg
12	100 m2	-	-	PI	0.15	1.20
15	100 m2	-	-	PI	0.15	1.50
20	100 m2	-	-	PI	0.15	2.00
25	100 m2	-	-	PI	0.15	2.50
30	100 m2	-	-	PI	0.15	3.00
35	100 m2	-	-	PI	0.15	3.50
40	100 m2	-	-	PI	0.15	4.00
50	100 m2	-	-	PI	0.15	5.00

	Unit	Labour grade	Labour hours	Plant grade	Plant hours	Materials kg

Supply and spread fertiliser to prepared ground (35g/m2)

	Unit	Labour grade	Labour hours	Plant grade	Plant hours	Materials kg
ammonium sulphate	100m2	-	-	PI	0.15	3.50
nitrochalk	100m2	-	-	PI	0.15	3.50
potassium nitrate	100m2	-	-	PI	0.15	3.50
urea	100m2	-	-	PI	0.15	3.50
super phosphate	100m2	-	-	PI	0.15	3.50
triple superphosphate	100m2	-	-	PI	0.15	3.50
potassium sulphate	100m2	-	-	PI	0.15	3.50
magnesium sulphate	100m2	-	-	PI	0.15	3.50
bonemeal	100m2	-	-	PI	0.15	3.50
dried blood	100m2	-	-	PI	0.15	3.50
hoof and horn	100m2	-	-	PI	0.15	3.50
fish, blood and stone	100m2	-	-	PI	0.15	3.50

Post-seeding treatment (by machine)

	Unit	Labour grade	Labour hours	Plant grade	Plant hours	Materials kg
Treat with light chain harrow	100m2	-	-	PI	0.03	-

	Unit	Labour grade	Labour hours	Plant grade	Plant hours

Turfing

Lay imported untreated meadow turf on prepared surfaces

	Unit	Labour grade	Labour hours	Plant grade	Plant hours
general areas	m2	LB	0.10	-	-
areas 10-45 degrees to the horizontal	m2	LB	0.12	-	-
areas exceeding 45 degrees to the horizontal	m2	LB	0.18	-	-

	Unit	Labour grade	Labour hours	Plant grade	Plant hours
Treat with wooden paddle beater	m2	LB	0.03	-	-
Treat turves with light roller	m2	LB	0.02	PJ	0.01

Trees and shrubs

Tree sizes

Specification	Height m	Clear stem height m	Girth cm
Light standard	2.50-2.75	1.5-1.8	6-8
Standard	2.75-3.00	1.8	8-10
Selected standard	3.00-3.50	1.8	10-12
Heavy standard	3.50-4.00	1.8	12-14
Extra heavy standard	4.00-5.00	1.8	12-14

Labour grades

Craftsman   LA

	Unit	Labour grade	Labour hours	Peat m3	Manure m3

Transplants and seedlings

Excavate for and plant transplants or seedlings, backfill and water

	Unit	Labour grade	Labour hours	Peat m3	Manure m3
20-40cm high	nr	LA	0.05	-	-
40-60cm high	nr	LA	0.05	-	-
60-90cm high	nr	LA	0.05	-	-
90-120cm high	nr	LA	0.05	-	-

Bare root trees

Excavate tree pit by hand, fork bottom of pit, plant containerised tree, backfill with excavated material including organic manure (30% of soil by volume), water and surround with peat

	Unit	Labour grade	Labour hours	Peat m3	Manure m3
6-8cm girth	nr	LA	1.50	0.060	0.015
8-10cm girth	nr	LA	2.00	0.060	0.015
10-12cm girth	nr	LA	2.00	0.070	0.020
12-14cm girth	nr	LA	2.50	0.070	0.020
14-16cm girth	nr	LA	2.50	0.075	0.020
16-18cm girth	nr	LA	3.00	0.075	0.025
18-20cm girth	nr	LA	4.00	0.075	0.025
20-25cm girth	nr	LA	5.00	0.080	0.025
120-150cm high	nr	LA	1.50	0.080	0.025
150-180cm high	nr	LA	2.00	0.075	0.025
180-240cm high	nr	LA	2.50	0.075	0.025
240-300cm high	nr	LA	4.00	0.080	0.025
300-350cm high	nr	LA	4.00	0.080	0.025

	Unit	Labour grade	Labour hours	Peat m3	Manure m3

Rootballed trees

Excavate tree pit by hand,
fork bottom of pit, plant
rootballed tree,
backfill with excavated
material including organic
manure (30% of soil by
volume), water and
surround with peat

	Unit	Labour grade	Labour hours	Peat m3	Manure m3
6-8cm girth	nr	LA	1.50	0.060	0.015
8-10cm girth	nr	LA	2.00	0.060	0.015
10-12cm girth	nr	LA	2.00	0.070	0.020
12-14cm girth	nr	LA	2.50	0.070	0.020
14-16cm girth	nr	LA	2.50	0.075	0.020
16-18cm girth	nr	LA	3.00	0.075	0.025
18-20cm girth	nr	LA	4.00	0.075	0.025
20-25cm girth	nr	LA	5.00	0.080	0.025
120-150cm high	nr	LA	1.50	0.080	0.025
150-180cm high	nr	LA	2.00	0.075	0.025
180-240cm high	nr	LA	2.50	0.075	0.025
240-300cm high	nr	LA	4.00	0.080	0.025
300-350cm high	nr	LA	4.00	0.080	0.025

	Unit	Labour grade	Labour hours	Peat m3	Manure m3

Conifers

Excavate tree pit by hand, fork bottom of pit, plant conifer, backfill with excavated material including organic manure (30% of soil by volume), water and surround with peat

	Unit	Labour grade	Labour hours	Peat m3	Manure m3
25-30cm high	nr	LA	0.05	0.020	0.005
30-40cm high	nr	LA	0.10	0.020	0.005
40-50cm high	nr	LA	0.15	0.025	0.010
50-60cm high	nr	LA	0.15	0.025	0.010
60-70cm high	nr	LA	0.20	0.030	0.015
70-80cm high	nr	LA	0.20	0.030	0.015

Shrubs

Form planting hole in cultivated area, place plant in hole, backfill with excavated material including organic manure (30% of soil by volume), water and surround with peat

	Unit	Labour grade	Labour hours	Peat m3	Manure m3
25-30cm high	nr	LA	0.05	0.020	0.005
30-40cm high	nr	LA	0.10	0.020	0.005
40-50cm high	nr	LA	0.15	0.025	0.010
50-60cm high	nr	LA	0.15	0.025	0.010
60-70cm high	nr	LA	0.20	0.030	0.015
70-80cm high	nr	LA	0.20	0.030	0.015
80-90cm high	nr	LA	0.25	0.030	0.015
90-120cm high	nr	LA	0.30	0.030	0.015

	Unit	Labour grade	Labour hours	Peat m3	Manure m3
Climbers					

Form planting hole, place
plant in hole, backfill with
excavated material including
organic manure (30% of soil
by volume), water and
surround with peat

	Unit	Labour grade	Labour hours	Peat m3	Manure m3
45-60cm high	nr	LA	0.05	0.030	0.015
50-60cm high	nr	LA	0.05	0.030	0.015

	Unit	Labour grade	Labour hours	Plants
Hedges				

Excavate trench size
300 x 300mm, lay soil aside
for future use, plant single
row of hedging plants and
carefully replace soil around
roots, plants at centres of

	Unit	Labour grade	Labour hours	Plants
150mm	m	LA	0.50	6.66
200mm	m	LA	0.55	5.00
250mm	m	LA	0.60	4.00
300mm	m	LA	0.65	3.33

Excavate trench size
300 x 300mm, lay soil aside
for future use, plant double
row of hedging plants and
carefully replace soil around
roots, plants at centres of

	Unit	Labour grade	Labour hours	Plants
250mm	m	LA	0.75	8.00
300mm	m	LA	0.80	6.66
350mm	m	LA	0.85	5.70
400mm	m	LA	0.90	5.00

	Unit	Labour grade	Labour hours	Bulbs

Bulbs

Prepare ground and
plant bulbs, density

	Unit	Labour grade	Labour hours	Bulbs
four per m2	m2	LA	0.30	4.00
six per m2	m2	LA	0.35	6.00
eight per m2	m2	LA	0.40	8.00
ten per m2	m2	LA	0.45	10.00
twelve per m2	m2	LA	0.50	12.00

	Unit	Labour grade	Labour hours	Plants

Groundcover plants

Prepare ground and
plant groundcover plants,
density

	Unit	Labour grade	Labour hours	Plants
four per m2	m2	LA	0.30	4.00
six per m2	m2	LA	0.35	6.00
eight per m2	m2	LA	0.40	8.00
ten per m2	m2	LA	0.45	10.00
twelve per m2	m2	LA	0.50	12.00

34

Land drainage

Weights of materials		kg/m3
Ashes		800
Gravel		1750
Limestone, crushed		1760
Sand		1600
		kg/m
PVC-U pipes,	80mm	1.20
	110mm	1.60
	160mm	3.00
	200mm	4.60
	250mm	7.20
Vitrified clay pipes,	100mm	15.63
	150mm	37.04
	225mm	95.24
	300mm	196.08
	400mm	357.14
	450mm	500.00
	500mm	555.60

Volumes of filling (m3/m)

Pipe dia. mm	Beds			Bed and haunching	Surround
	50mm	100mm	150mm		
75	0.017	0.034	0.051	0.088	0.139
100	0.023	0.045	0.068	0.117	0.185
150	0.026	0.053	0.079	0.152	0.231
225	0.030	0.060	0.090	0.195	0.285

Trench widths

Pipe dia. mm	Less than 1.5m deep mm	More than 1.5m deep mm
75	450	600
100	450	600
150	500	650
225	600	750
300	650	800
400	750	900
450	900	1050
600	1000	1300

Labour grades

Semi-skilled operative	LB
Unskilled operative	LC

Plant grades

Hydraulic excavator (1.7m3)	PA
Land drain trencher	PZA

	Unit	Labour grade	Labour hours	Plant grade	Plant hours	Trench volume m3

Machine excavation by trencher

Excavate trench 150mm wide for drain, remove excavated material, average depth

0.50m	m	-	-	PZA	0.05	0.075
0.75m	m	-	-	PZA	0.08	0.113

Excavate trench 250mm wide for drain, remove excavated material, average depth

0.50m	m	-	-	PZA	0.08	0.125
0.75m	m	-	-	PZA	0.10	0.185
1.00m	m	-	-	PZA	0.12	0.250

Excavate trench 300mm wide for drain, remove excavated material, average depth

0.50m	m	-	-	PZA	0.10	0.150
0.75m	m	-	-	PZA	0.12	0.255
1.00m	m	-	-	PZA	0.14	0.300

Excavate trench 400mm wide for drain, remove excavated material, average depth

0.50m	m	-	-	PZA	0.12	0.200
0.75m	m	-	-	PZA	0.14	0.300
1.00m	m	-	-	PZA	0.16	0.400

	Unit	Labour grade	Labour hours	Plant grade	Plant hours	Filling m3
Gravel rejects in trench 150mm wide						
0.50m depth	m	LC	0.11	-	-	0.075
0.75m depth	m	LC	0.17	-	-	0.113
Gravel rejects in trench 250mm wide						
0.50m depth	m	LC	0.15	-	-	0.150
0.75m depth	m	LC	0.23	-	-	0.225
1.00m depth	m	LC	0.30	-	-	0.300
Gravel rejects in trench 300mm wide						
0.50m depth	m	LC	0.15	-	-	0.150
0.75m depth	m	LC	0.23	-	-	0.225
1.00m depth	m	LC	0.30	-	-	0.300
Gravel rejects in trench 400mm wide						
0.50m depth	m	LC	0.15	-	-	0.200
0.75m depth	m	LC	0.23	-	-	0.300
1.00m depth	m	LC	0.30	-	-	0.400

	Unit	Labour grade	Labour hours	Pipes nr
Pipework				
Agricultural clay field drain pipes, 300mm long laid butt jointed in trench, diameter				
75mm	m	LB	0.12	3.33
100mm	m	LB	0.14	3.33
150mm	m	LB	0.16	3.33
225mm	m	LB	0.20	3.33

	Unit	Labour grade	Labour hours	Pipes nr
Single junction, diameter				
75mm	nr	LB	0.12	1.00
100mm	nr	LB	0.14	1.00
150mm	nr	LB	0.16	1.00
PVC-U plain ended field drain pipes, laid butt jointed in trench, diameter				
110mm	m	LB	0.16	1.00
160mm	m	LB	0.18	1.00
Couplers, diameter				
110mm	nr	LB	0.16	1.00
160mm	nr	LB	0.18	1.00
Bends, short radius, diameter				
110mm	nr	LB	0.16	1.00
160mm	nr	LB	0.18	1.00
Vitrified clay perforated drain pipes, laid butt jointed in trench, diameter				
100mm	m	LB	0.16	1.00
150mm	m	LB	0.18	1.00
225mm	m	LB	0.20	1.00
Bends, diameter				
100mm	nr	LB	0.16	1.00
150mm	nr	LB	0.18	1.00
225mm	nr	LB	0.20	1.00
Junctions, diameter				
75mm	nr	LB	0.16	1.00
100mm	nr	LB	0.18	1.00
150mm	nr	LB	0.20	1.00

	Unit	Labour grade	Labour hours	Plant grade	Plant hours	Trench m3

Ditching

Form ditch with 45 degree sides by machine average width 500mm, depth

	Unit	Labour grade	Labour hours	Plant grade	Plant hours	Trench m3
500mm	m	-	-	PA	0.10	0.125
1000mm	m	-	-	PA	0.14	0.250
1500mm	m	-	-	PA	0.18	0.375
2000mm	m	-	-	PA	0.22	0.500

Form ditch with 45 degree sides by machine average width 1000mm, depth

	Unit	Labour grade	Labour hours	Plant grade	Plant hours	Trench m3
500mm	m	-	-	PA	0.14	0.250
1000mm	m	-	-	PA	0.22	0.500
1500mm	m	-	-	PA	0.34	0.750
2000mm	m	-	-	PA	0.46	1.000

Form ditch with 45 degree sides by machine average width 1500mm, depth

	Unit	Labour grade	Labour hours	Plant grade	Plant hours	Trench m3
500mm	m	-	-	PA	0.18	0.375
1000mm	m	-	-	PA	0.34	0.750
1500mm	m	-	-	PA	0.50	1.125
2000mm	m	-	-	PA	0.70	1.500

35

Water supply and ponds

Pond lining materials	Sheet size m
Butyl	4 x 3
	5 x 4
	6 x 4
PVC-U	3 x 2
	4 x 3
	5 x 4
	6 x 6
Heavy duty polyolefin	3 x 2
	4 x 3
	6 x 4
	7 x 6

Labour grades

Craftsman	LA
Semi-skilled operative	LB

	Unit	Labour grade	Labour hours
Pipework			
Medium density polyethylene pipe (MDPE), diameter			
20mm	m	LA	0.11
25mm	m	LA	0.11
32mm	m	LA	0.17
50mm	m	LA	0.17
63mm	m	LA	0.21
Gunmetal stopcock for underground use, dezincification resistant, polyethylene x polyethylene			
20mm	nr	LA	0.22
25mm	nr	LA	0.25
Hose union bib tap including back plate plugged and screwed to brickwork			
15mm	nr	LA	0.15
Pond liners			
Liner laid in prepared excavation			
Butyl	m2	LB	0.02
PVC-U	m2	LB	0.02
Heavy duty polyolefin	m2	LB	0.02

	Unit	Labour grade	Labour hours

Semi-rigid pools

Semi-rigid pools placed
in prepared excavation
including backfilling with
sand around irregular
contours of pool

	Unit	Labour grade	Labour hours
1750 x 1250 x 500mm (300 litres)	nr	LB	0.50
2000 x 1500 x 500mm (480 litres)	nr	LB	0.70
2500 x 1450 x 500mm (600 litres)	nr	LB	0.80

Pumps

	Unit	Labour grade	Labour hours
Pond pumps, 65 gallons per hour, delivery height 700mm	nr	LB	1.00
Pond pumps, 130 gallons per hour, delivery height 1150mm	nr	LB	1.00
Pond pumps, 650 gallons per hour, delivery height 3250mm	nr	LB	1.00

PART FOUR

MECHANICAL WORK

36

Piped supply systems

Size of expansion tanks

Boiler rating kW	Tank size litres	BS Ref.
12	54	SCM 90
25	30	SCM 90
30	68	SCM 110
45	68	SCM 110
55	86	SCM 135
75	114	SCM 180
150	191	SCM 270
225	227	SCM 320
275	264	SCM 360
375	327	SCM 450/1
400	336	SCM 450/2
550	423	SCM 570
800	709	SCM 910
900	841	SCM 1600
1200	1227	SCM 1600

BS Ref.	Ball valve nominal bore size mm	Cold feed nominal bore size mm	Open vent nominal bore size mm	Overflow nominal bore size mm
SCM 90	15	20	25	25
SCM 90	15	20	25	32
SCM 110	15	20	25	32
SCM 110	15	20	25	32
SCM 135	15	20	25	32
SCM 180	15	25	32	32
SCM 270	15	25	32	32
SCM 320	20	32	40	40
SCM 360	20	32	40	40
SCM 450/1	20	40	50	40
SCM 450/2	20	40	50	40
SCM 570	25	40	50	50
SCM 910	25	50	65	50
SCM 1130	25	50	65	65
SCM 1600	25	50	65	65

Copper pipe, table Z

Size of pipe mm	Outside diameter maximum mm	minimum mm	Nominal thickness mm	Maximum working pressures bars
6	6.045	5.965	0.5	113
8	8.045	7.965	0.5	98
10	10.045	9.965	0.5	78
12	12.045	11.965	0.5	64
15	15.045	14.965	0.5	50
18	18.045	17.965	0.6	50
22	22.045	21.965	0.6	41
28	28.045	27.965	0.6	32
35	35.070	34.990	0.7	30
42	42.070	41.990	0.8	28
54	54.070	53.990	0.9	25
67	66.750	66.600	1.0	20
108	108.250	108.000	1.2	17
133	133.500	133.250	1.5	16

Copper pipe, Table X

Size of pipe mm	Outside diameter maximum mm	minimum mm	Nominal thickness mm	Maximum working pressures bars
6	6.045	5.965	0.6	113
8	8.045	7.965	0.6	98
10	10.045	9.965	0.6	78
12	12.045	11.965	0.6	64
15	15.045	14.965	0.7	50
18	18.045	17.965	0.7	50
22	22.045	21.965	0.8	41
28	28.045	27.965	0.9	32
35	35.070	34.990	1.2	30
42	42.070	41.990	1.2	28
54	54.070	53.990	1.2	25
67	66.750	66.600	1.2	20
108	108.250	108.000	1.5	17
133	133.500	133.250	1.5	16

Labour grades

1 Foreman, 1 advanced fitter/welder
 (gas/arc), 2 advanced fitters (gas or arc),
3 advanced fitters, 2 fitters and 1 mate   LP

	Unit	Labour grade	Labour hours

Copper pipelines

Copper pipes, capillary fittings,
including standard supports, diameter

	Unit	Labour grade	Labour hours
15mm	m	LP	0.16
extra for made bend	nr	LP	0.12
extra for stop end	nr	LP	0.15
extra for straight coupling	nr	LP	0.29
extra for reducing coupling	nr	LP	0.29
extra for male connector	nr	LP	0.29
extra for elbow	nr	LP	0.29
extra for tee	nr	LP	0.44
extra for tap connector	nr	LP	0.24
22mm	m	LP	0.23
extra for made bend	nr	LP	0.19
extra for stop end	nr	LP	0.21
extra for straight coupling	nr	LP	0.41
extra for reducing coupling	nr	LP	0.41
extra for male connector	nr	LP	0.41
extra for elbow	nr	LP	0.41
extra for tee	nr	LP	0.62
extra for tap connector	nr	LP	0.28

	Unit	Labour grade	Labour hours
28mm	m	LP	0.30
extra for made bend	nr	LP	0.29
extra for stop end	nr	LP	0.25
extra for straight coupling	nr	LP	0.50
extra for reducing coupling	nr	LP	0.50
extra for male connector	nr	LP	0.50
extra for elbow	nr	LP	0.50
extra for tee	nr	LP	0.75
35mm	m	LP	0.37
extra for made bend	nr	LP	0.40
extra for stop end	nr	LP	0.27
extra for straight coupling	nr	LP	0.53
extra for reducing coupling	nr	LP	0.53
extra for male connector	nr	LP	0.53
extra for elbow	nr	LP	0.53
extra for tee	nr	LP	0.79
extra for tank connector	nr	LP	0.40
42mm	m	LP	0.44
extra for made bend	nr	LP	0.51
extra for stop end	nr	LP	0.28
extra for straight coupling	nr	LP	0.55
extra for reducing coupling	nr	LP	0.55
extra for male connector	nr	LP	0.55
extra for elbow	nr	LP	0.55
extra for tee	nr	LP	0.84
extra for tank connector	nr	LP	0.46

	Unit	Labour grade	Labour hours
54mm	m	LP	0.57
extra for made bend	nr	LP	0.70
extra for stop end	nr	LP	0.29
extra for straight coupling	nr	LP	0.59
extra for reducing coupling	nr	LP	0.59
extra for male connector	nr	LP	0.59
extra for elbow	nr	LP	0.59
extra for tee	nr	LP	0.88
extra for tank connector	nr	LP	0.49
67mm	m	LP	0.73
extra for stop end	nr	LP	0.44
extra for straight coupling	nr	LP	0.88
extra for reducing coupling	nr	LP	0.88
extra for male connector	nr	LP	0.88
extra for elbow	nr	LP	0.88
extra for tee	nr	LP	1.32
extra for tank connector	nr	LP	0.70
76mm pipes	m	LP	0.87
extra for stop end	nr	LP	0.56
extra for straight coupling	nr	LP	1.12
cxtra for reducing coupling	nr	LP	1.12
extra for elbow	nr	LP	1.12
extra for tee	nr	LP	1.68
extra for tank connector	nr	LP	0.87

	Unit	Labour grade	Labour hours
108mm	m	LP	1.12
extra for stop end	nr	LP	0.57
extra for straight coupling	nr	LP	1.15
extra for reducing coupling	nr	LP	1.15
extra for elbow	nr	LP	1.15
extra for tee	nr	LP	1.72
extra for tank connector	nr	LP	1.03

Copper pipes, DZR non-manipulative compression fittings, including standard supports, diameter

	Unit	Labour grade	Labour hours
15mm	m	LP	0.06
extra for made bend	nr	LP	0.12
extra for stop end	nr	LP	0.06
extra for straight coupling	nr	LP	0.12
extra for elbow	nr	LP	0.12
extra for tee	nr	LP	0.18
extra for tank connector	nr	LP	0.15
22mm	m	LP	0.23
extra for made bend	nr	LP	0.19
extra for stop end	nr	LP	0.12
extra for straight coupling	nr	LP	0.24
extra for elbow	nr	LP	0.24
extra for tee	nr	LP	0.35
extra for tank connector	nr	LP	0.19

	Unit	Labour grade	Labour hours
28mm	m	LP	0.30
extra for made bend	nr	LP	0.29
extra for stop end	nr	LP	0.16
extra for straight coupling	nr	LP	0.32
extra for elbow	nr	LP	0.32
extra for tee	nr	LP	0.49
extra for tank connector	nr	LP	0.26
35mm	m	LP	0.37
extra for made bend	nr	LP	0.40
extra for stop end	nr	LP	0.18
extra for straight coupling	nr	LP	0.35
extra for elbow	nr	LP	0.35
extra for tee	nr	LP	0.53
extra for tank connector	nr	LP	0.31
42mm pipes	m	LP	0.44
extra for made bend	nr	LP	0.51
extra for stop end	nr	LP	0.19
extra for straight coupling	nr	LP	0.37
extra for elbow	nr	LP	0.37
extra for tee	nr	LP	0.57
extra for tank connector	nr	LP	0.38

Flanged joint, bimetal, comprising 1
flange, 1 corrugated brass joint ring
with nuts, bolts and washers, diameter

	Unit	Labour grade	Labour hours
67mm	nr	LP	0.52
76mm	nr	LP	0.52
108mm	nr	LP	0.59

	Unit	Labour grade	Labour hours
Flanged joint, bimetal, comprising 2 flanges, 1 corrugated brass joint ring with nuts, bolts and washers, diameter			
67mm	nr	LP	0.88
76mm	nr	LP	0.88
108mm	nr	LP	1.03
Flanged joint, brazing metal slip-on comprising 1 flange, 1 corrugated brass joint ring with nuts, bolts and washers, diameter			
67mm	nr	LP	0.52
76mm	nr	LP	0.52
108mm	nr	LP	0.59
Flanged joint, brazing metal slip-on comprising 2 flanges, 1 corrugated brass joint ring with nuts, bolts and washers, diameter			
67mm	nr	LP	0.88
76mm	nr	LP	0.88
108mm	nr	LP	1.03
Flanged joint, gunmetal, comprising 1 blank flange, 1 corrugated brass joint ring, with nuts, bolts and washers, diameter			
67mm	nr	LP	0.52
76mm	nr	LP	0.52
108mm	nr	LP	0.59

	Unit	Labour grade	Labour hours

Pipeline supports

Copper saddle band, fixing to
background, for pipe diameter

15mm	nr	LP	0.10
22mm	nr	LP	0.10
28mm	nr	LP	0.10
35mm	nr	LP	0.10
42mm	nr	LP	0.10
54mm	nr	LP	0.10

Copper single spacing clip, fixing to
background, for pipe diameter

15mm	nr	LP	0.10
22mm	nr	LP	0.10
28mm	nr	LP	0.10

Copper two-piece spacing clip, fixing to
background, for pipe diameter

15mm	nr	LP	0.15
22mm	nr	LP	0.15
28mm	nr	LP	0.15
35mm	nr	LP	0.15
42mm	nr	LP	0.15
54mm	nr	LP	0.15

	Unit	Labour grade	Labour hours
Brass pipe bracket, fixing to background, for pipe diameter			
15mm	nr	LP	0.10
22mm	nr	LP	0.10
28mm	nr	LP	0.10
Brass wall bracket, round baseplate, fixing to background, for pipe diameter			
15mm	nr	LP	0.10
22mm	nr	LP	0.10
28mm	nr	LP	0.10
35mm	nr	LP	0.10
42mm	nr	LP	0.10
54mm	nr	LP	0.10
Hospital bracket, fixing to background, for pipe diameter			
15mm	nr	LP	0.10
22mm	nr	LP	0.10
28mm	nr	LP	0.10
35mm	nr	LP	0.10
42mm	nr	LP	0.10
54mm	nr	LP	0.10

	Unit	Labour grade	Labour hours

Pipe support for single pipe comprising
threaded brass stem, female backplate
250mm long suspended from soffit,
fixing to background, for pipe diameter

	Unit	Labour grade	Labour hours
15mm	nr	LP	0.15
22mm	nr	LP	0.15
28mm	nr	LP	0.15
35mm	nr	LP	0.15
42mm	nr	LP	0.15
54mm	nr	LP	0.15

Copper wall/floor sleeves 250mm long,
formed from offcuts, for copper pipe,
diameter

	Unit	Labour grade	Labour hours
15mm	nr	LP	0.03
22mm	nr	LP	0.03
28mm	nr	LP	0.04
35mm	nr	LP	0.06
42mm	nr	LP	0.07
54mm	nr	LP	0.09
67mm	nr	LP	0.15
76mm	nr	LP	0.20
108mm	nr	LP	0.25

	Unit	Labour grade	Labour hours

Copper wall/floor sleeves 250-500mm long, formed from offcuts, for copper pipe, diameter

	Unit	Labour grade	Labour hours
15mm	nr	LP	0.03
22mm	nr	LP	0.03
28mm	nr	LP	0.04
35mm	nr	LP	0.06
42mm	nr	LP	0.07
54mm	nr	LP	0.09
67mm	nr	LP	0.15
76mm	nr	LP	0.20
108mm	nr	LP	0.25

Wall/floor cover plates

Plastic, for copper pipe, fixing to pipes diameter

	Unit	Labour grade	Labour hours
15mm	nr	LP	0.10
22mm	nr	LP	0.10
28mm	nr	LP	0.10
35mm	nr	LP	0.10
42mm	nr	LP	0.10
54mm	nr	LP	0.10

Pipework ancillaries

Gunmetal stopcock with crutch head, diameter

	Unit	Labour grade	Labour hours
15mm	nr	LP	0.58
22mm	nr	LP	0.68
28mm	nr	LP	0.74

	Unit	Labour grade	Labour hours
Gunmetal lockshield stopcock, diameter			
15mm	nr	LP	0.58
22mm	nr	LP	0.68
28mm	nr	LP	0.74
Non-dezincifiable alloy stopcock with crutch head, diameter			
15mm	nr	LP	0.13
22mm	nr	LP	0.26
28mm	nr	LP	0.36
Non-dezincifiable alloy lockshield stopcock, diameter			
15mm	nr	LP	0.13
22mm	nr	LP	0.26
28mm	nr	LP	0.36
Gunmetal gate valve with wheel head and capillary end, diameter			
15mm	nr	LP	0.52
22mm	nr	LP	0.58
28mm	nr	LP	0.58
35mm	nr	LP	0.65
42mm	nr	LP	0.79
54mm	nr	LP	0.80

	Unit	Labour grade	Labour hours
Sprinklers			
Mild steel pipes with black malleable iron fittings excluding supports, prefabricated, nominal bore			
20mm	m	LP	0.20
erect prefabricated length	m	LP	0.85
make screwed joint	nr	LP	0.07
25mm	m	LP	0.26
erect prefabricated length	m	LP	0.93
make screwed joint	nr	LP	0.12
32mm	m	LP	0.27
erect prefabricated length	m	LP	1.00
make screwed joint	nr	LP	0.13
40mm	m	LP	0.29
erect prefabricated length	m	LP	1.08
make screwed joint	nr	LP	0.19
50mm	m	LP	0.32
erect prefabricated length	m	LP	1.15
make screwed joint	nr	LP	0.20

	Unit	Labour grade	Labour hours
65mm	m	LP	0.39
erect prefabricated length	m	LP	1.23
make screwed joint	nr	LP	0.27
make victuallic joint	nr	LP	0.69
80mm	m	LP	0.49
erect pre-fabricated length	m	LP	1.39
make screwed joint	nr	LP	0.32
make victuallic joint	nr	LP	0.69
100mm	m	LP	0.77
erect pre-fabricated length	m	LP	1.70
make screwed joint	nr	LP	0.47
make victuallic joint	nr	LP	0.77
125mm	m	LP	0.85
erect pre-fabricated length	m	LP	1.90
make screwed joint	nr	LP	0.48
make victuallic joint	nr	LP	0.93
150mm	m	LP	0.96
erect pre-fabricated length	m	LP	2.16
make screwed joint	nr	LP	0.60
make victuallic joint	nr	LP	0.93

	Unit	Labour grade	Labour hours
Sprinkler heads, diameter			
15mm	nr	LP	0.25
20mm	nr	LP	0.25
Multi-jet			
20mm	nr	LP	0.07
25mm	nr	LP	0.12
Sprinkler guards, diameter			
15mm	nr	LP	0.05
Sprinkler rosettes, diameter			
15mm	nr	LP	0.05
Control valves with flanged ends			
100mm	nr	LP	1.54
150mm	nr	LP	2.70
200mm	nr	LP	3.47
Subsidiary air valves			
100mm	nr	LP	1.54
150mm	nr	LP	2.70
200mm	nr	LP	3.47
Alarm gong			
internal type	nr	LP	3.86
external type	nr	LP	1.54

	Unit	Labour grade	Labour hours
Air compressor	nr	LP	1.16
Jockey pump	nr	LP	1.16
Valve lock and strap	nr	LP	0.08
Temporary plugging of sprinkler outlet, diameter			
15mm	nr	LP	0.05
20mm	nr	LP	0.07
25mm	nr	LP	0.12
32mm	nr	LP	0.13

37

Heating and cooling systems

Heat calculations

Heat lost from a building can be expressed as the product of the area of the surface, its thermal transmission coefficient and the difference between the outside and inside temperatures, viz.

$$Q = AU(T1 - T2)$$

where Q is the heat lost through the fabric of the building in watts, A is the area of the surface of the building in square metres, U is the thermal transmission coefficient W/m2 degrees centigrade, T1 is the inside temperature required and T2 is the outside temperature. An allowance must also be made for the heat lost through floors and ceilings. The power required can be assessed in approximate terms by allowing 49 watts per square metre to produce a temperature of 20 degrees centigrade with an outside temperature of 0 degrees centigrade.

Capacity of tanks and cylinders

Rectangular container gallons = length x width x height divided by 4546

Rectangular container litres = length x width x height divided by 1000

Cylindrical container gallons = radius x radius x height divided by 5788

Cylindrical container litres = radius x radius x height divided by 1273

(All above dimensions to be in centimetres)

Power required

Minutes required to heat water = litres x temperature rise in degrees centigrade divided by 14.33 x kW x efficiency

Loading required in kW to heat water = litres x temperature rise in degrees centigrade divided by 14.33 x minutes x efficiency

Labour gangs

1 Foreman, 1 advance fitter/welder
(gas/arc), 2 advanced fitters (gas or arc)
3 advanced fitters, 2 fitters and 1 mate   LP

	Unit	Labour grade	Labour hours

Mild steel pipework, screwed joints

Mild steel pipes with black malleable
iron fittings excluding supports,
nominal bore

	Unit	Labour grade	Labour hours
15mm	m	LP	0.22
extra for made bend	nr	LP	0.12
extra for made offset	nr	LP	0.27
extra for plug	nr	LP	0.24
extra for cap	nr	LP	0.24
extra for nipple	nr	LP	0.24
extra for union	nr	LP	0.61
extra for reducer	nr	LP	0.49
extra for elbow, 90 degrees	nr	LP	0.49
extra for equal tee	nr	LP	0.73
20mm	m	LP	0.24
extra for made bend	nr	LP	0.26
extra for made offset	nr	LP	0.44
extra for plug	nr	LP	0.26
extra for cap	nr	LP	0.26
extra for nipple	nr	LP	0.26
extra for union	nr	LP	0.66
extra for reducer	nr	LP	0.52
extra for elbow, 90 degrees	nr	LP	0.52
extra for equal tee	nr	LP	0.78

	Unit	Labour grade	Labour hours
25mm	m	LP	0.30
extra for made bend	nr	LP	0.38
extra for made offset	nr	LP	0.66
extra for plug	nr	LP	0.29
extra for cap	nr	LP	0.29
extra for nipple	nr	LP	0.29
extra for union	nr	LP	0.73
extra for reducer	nr	LP	0.58
extra for elbow, 90 degrees	nr	LP	0.58
extra for equal tee	nr	LP	0.87
32mm	m	LP	0.36
extra for made bend	nr	LP	0.50
extra for made offset	nr	LP	0.90
extra for plug	nr	LP	0.32
extra for cap	nr	LP	0.32
extra for nipple	nr	LP	0.32
extra for union	nr	LP	0.81
extra for reducer	nr	LP	0.65
extra for elbow, 90 degrees	nr	LP	0.65
extra for equal tee	nr	LP	0.97
40mm	m	LP	0.45
extra for made bend	nr	LP	0.65
extra for made offset	nr	LP	1.03
extra for plug	nr	LP	0.37
extra for cap	nr	LP	0.37
extra for nipple	nr	LP	0.37
extra for union	nr	LP	0.94
extra for reducer	nr	LP	0.74
extra for elbow, 90 degrees	nr	LP	0.74
extra for equal tee	nr	LP	1.02

	Unit	Labour grade	Labour hours
50mm	m	LP	0.61
extra for made bend	nr	LP	0.91
extra for made offset	nr	LP	1.37
extra for plug	nr	LP	0.40
extra for cap	nr	LP	0.40
extra for nipple	nr	LP	0.40
extra for union	nr	LP	1.02
extra for reducer	nr	LP	0.81
extra for elbow, 90 degrees	nr	LP	0.81
extra for equal tee	nr	LP	1.21
65mm	m	LP	0.74
extra for plug	nr	LP	0.49
extra for cap	nr	LP	0.49
extra for nipple	nr	LP	0.49
extra for union	nr	LP	1.21
extra for reducer	nr	LP	0.97
extra for elbow, 90 degrees	nr	LP	0.97
extra for equal tee	nr	LP	1.46
80mm	m	LP	0.90
extra for plug	nr	LP	0.55
extra for cap	nr	LP	0.55
extra for nipple	nr	LP	0.55
extra for union	nr	LP	1.37
extra for reducer	nr	LP	1.10
extra for elbow, 90 degrees	nr	LP	1.10
extra for equal tee	nr	LP	1.65

	Unit	Labour grade	Labour hours
100mm	m	LP	1.18
extra for plug	nr	LP	0.61
extra for cap	nr	LP	0.61
extra for nipple	nr	LP	0.61
extra for union	nr	LP	1.54
extra for reducer	nr	LP	1.21
extra for elbow, 90 degrees	nr	LP	1.21
extra for equal tee	nr	LP	1.84
125mm	m	LP	1.25
extra for plug	nr	LP	0.68
extra for cap	nr	LP	0.68
extra for nipple	nr	LP	0.68
extra for union	nr	LP	1.70
extra for reducer	nr	LP	1.36
extra for elbow, 90 degrees	nr	LP	1.36
extra for equal tee	nr	LP	1.84
150mm	m	LP	1.75
extra for plug	nr	LP	0.78
extra for cap	nr	LP	0.78
extra for nipple	nr	LP	0.78
extra for union	nr	LP	1.94
extra for reducer	nr	LP	1.55
extra for elbow, 90 degrees	nr	LP	1.55
extra for equal tee	nr	LP	2.33

	Unit	Labour grade	Labour hours
Carbon steel pipework and fittings, hot finished seamless, excluding supports, nominal bore			
25mm, 3.4mm thick	m	LP	0.33
extra for elbow, 45 degrees	nr	LP	0.58
extra for elbow, 90 degrees	nr	LP	0.58
extra for branch bend	nr	LP	0.58
extra for equal tee	nr	LP	0.87
32mm, 3.2mm thick	m	LP	0.39
extra for elbow, 45 degrees	nr	LP	0.65
extra for elbow, 90 degrees	nr	LP	0.65
extra for branch bend	nr	LP	0.65
extra for equal tee	nr	LP	0.97
40mm, 4.0mm thick	m	LP	0.49
extra for elbow, 45 degrees	nr	LP	0.74
extra for elbow, 90 degrees	nr	LP	0.74
extra for branch bend	nr	LP	0.74
extra for equal tee	nr	LP	1.02
50mm, 4.0mm thick	m	LP	0.65
extra for elbow, 45 degrees	nr	LP	0.81
extra for elbow, 90 degrees	nr	LP	0.81
extra for branch bend	nr	LP	0.81
extra for equal tee	nr	LP	1.21

	Unit	Labour grade	Labour hours
65mm, 5.00mm thick	m	LP	0.80
extra for elbow, 45 degrees	nr	LP	0.97
extra for elbow, 90 degrees	nr	LP	0.97
extra for branch bend	nr	LP	0.97
extra for equal tee	nr	LP	1.46
80mm, 4.5mm thick	m	LP	0.98
extra for elbow, 45 degrees	nr	LP	1.10
extra for elbow, 90 degrees	nr	LP	1.10
extra for branch bend	nr	LP	1.10
extra for equal tee	nr	LP	1.65
100mm, 6.3mm thick	m	LP	1.29
extra for elbow, 45 degrees	nr	LP	1.21
extra for elbow, 90 degrees	nr	LP	1.21
extra for branch bend	nr	LP	1.21
extra for equal tee	nr	LP	1.84
125mm, 6.3mm thick	m	LP	1.35
extra for elbow, 45 degrees	nr	LP	1.21
extra for elbow, 90 degrees	nr	LP	1.21
extra for branch bend	nr	LP	1.21
extra for equal tee	nr	LP	1.84
150mm, 5.6mm thick	m	LP	1.90
extra for elbow, 45 degrees	nr	LP	1.55
extra for elbow, 90 degrees	nr	LP	1.55
extra for branch bend	nr	LP	1.55
extra for equal tee	nr	LP	2.33

	Unit	Labour grade	Labour hours
200mm, 5.6mm thick	m	LP	2.48
extra for elbow, 45 degrees	nr	LP	1.94
extra for elbow, 90 degrees	nr	LP	1.94
extra for branch bend	nr	LP	1.94
extra for equal tee	nr	LP	2.90
250mm, 10.0mm thick	m	LP	2.86
extra for elbow, 45 degrees	nr	LP	2.43
extra for elbow, 90 degrees	nr	LP	2.43
extra for branch bend	nr	LP	2.43
extra for equal tee	nr	LP	3.64

Flanged joint, blank, comprising 1
flange, 1 corrugated brass joint ring
with nuts, bolts and washers, diameter

	Unit	Labour grade	Labour hours
15mm	nr	LP	0.24
20mm	nr	LP	0.26
25mm	m	LP	0.29
32mm	nr	LP	0.32
40mm	nr	LP	0.37
50mm	nr	LP	0.40
65mm	nr	LP	0.49
80mm	nr	LP	0.55
100mm	nr	LP	0.61
125mm	nr	LP	0.68
150mm	nr	LP	0.77
200mm	nr	LP	0.97

	Unit	Labour grade	Labour hours
Flanged joint, slip- on comprising 1 flange, 1 corrugated brass joint ring with nuts, bolts and washers, diameter			
15mm	nr	LP	0.29
20mm	nr	LP	0.34
25mm	nr	LP	0.37
32mm	nr	LP	0.42
40mm	nr	LP	0.50
50mm	nr	LP	0.65
65mm	nr	LP	0.71
80mm	nr	LP	0.84
100mm	nr	LP	1.02
125mm	nr	LP	1.37
150mm	nr	LP	1.62
200mm	nr	LP	2.02
Cover plates for steel pipes, clamped to pipe, diameter			
15mm	nr	LP	0.18
20mm	nr	LP	0.18
25mm	nr	LP	0.18
32mm	nr	LP	0.18
40mm	nr	LP	0.18
50mm	nr	LP	0.18

	Unit	Labour grade	Labour hours

Mild steel pipework, gas welding

Mild steel pipes with black heavy malleable iron fittings excluding supports, (flange welds include for two outer and one inner welds), nominal bore

	Unit	Labour grade	Labour hours
15mm	m	LP	0.14
extra for butt weld	nr	LP	0.25
extra for square branch weld	nr	LP	0.42
extra for swept branch weld	nr	LP	0.68
extra for flange weld	nr	LP	0.32
20mm	m	LP	0.16
extra for butt weld	nr	LP	0.36
extra for square branch weld	nr	LP	0.62
extra for swept branch weld	nr	LP	0.88
extra for flange weld	nr	LP	0.46
25mm	m	LP	0.19
extra for butt weld	nr	LP	0.37
extra for square branch weld	nr	LP	0.84
extra for swept branch weld	nr	LP	1.11
extra for flange weld	nr	LP	0.56
32mm	m	LP	0.23
extra for butt weld	nr	LP	0.48
extra for square branch weld	nr	LP	1.02
extra for swept branch weld	nr	LP	1.28
extra for flange weld	nr	LP	0.62

	Unit	Labour grade	Labour hours
40mm	m	LP	0.29
extra for butt weld	nr	LP	0.57
extra for square branch weld	nr	LP	1.23
extra for swept branch weld	nr	LP	1.50
extra for flange weld	nr	LP	0.75
50mm	m	LP	0.39
extra for butt weld	nr	LP	0.75
extra for square branch weld	nr	LP	1.62
extra for swept branch weld	nr	LP	1.87
extra for flange weld	nr	LP	0.97
65mm	m	LP	0.47
extra for butt weld	nr	LP	0.93
extra for square branch weld	nr	LP	2.02
extra for swept branch weld	nr	LP	2.28
extra for flange weld	nr	LP	1.22

Mild steel pipework, arc welding

Mild steel pipes with black heavy malleable iron fittings excluding supports, (flange welds include for two outer and one inner welds), nominal bore

	Unit	Labour grade	Labour hours
80mm	m	LP	0.58
extra for butt weld	nr	LP	1.07
extra for square branch weld	nr	LP	1.94
extra for swept branch weld	nr	LP	2.25
extra for flange weld	nr	LP	1.39

	Unit	Labour grade	Labour hours
100mm	m	LP	0.76
extra for butt weld	nr	LP	1.39
extra for square branch weld	nr	LP	2.08
extra for swept branch weld	nr	LP	2.39
extra for flange weld	nr	LP	1.70
125mm	m	LP	0.79
extra for butt weld	nr	LP	1.70
extra for square branch weld	nr	LP	2.93
extra for swept branch weld	nr	LP	3.24
extra for flange weld	nr	LP	2.22
150mm	m	LP	1.12
extra for butt weld	nr	LP	2.00
extra for square branch weld	nr	LP	3.40
extra for swept branch weld	nr	LP	3.24
extra for flange weld	nr	LP	2.62
200mm	m	LP	1.46
extra for butt weld	nr	LP	2.65
extra for square branch weld	nr	LP	4.32
extra for swept branch weld	nr	LP	4.78
extra for flange weld	nr	LP	3.47
250mm	m	LP	1.80
extra for butt weld	nr	LP	4.54
extra for square branch weld	nr	LP	5.57
extra for swept branch weld	nr	LP	6.34
extra for flange weld	nr	LP	4.85

	Unit	Labour grade	Labour hours

Mild steel prefabricated pipework, gas welding

Mild steel pipes with black heavy malleable iron fittings excluding supports, (flange welds include for two outer and one inner welds), nominal bore

	Unit	Labour grade	Labour hours
15mm	m	LP	0.14
extra for butt weld	nr	LP	0.17
extra for square branch weld	nr	LP	0.31
extra for swept branch weld	nr	LP	0.59
extra for flange weld	nr	LP	0.31
20mm	m	LP	0.16
extra for butt weld	nr	LP	0.28
extra for square branch weld	nr	LP	0.42
extra for swept branch weld	nr	LP	0.74
extra for flange weld	nr	LP	0.40
25mm	m	LP	0.19
extra for butt weld	nr	LP	0.34
extra for square branch weld	nr	LP	0.49
extra for swept branch weld	nr	LP	0.88
extra for flange weld	nr	LP	0.49
32mm	m	LP	0.23
extra for butt weld	nr	LP	0.42
extra for square branch weld	nr	LP	0.59
extra for swept branch weld	nr	LP	1.00
extra for flange weld	nr	LP	0.57

	Unit	Labour grade	Labour hours
40mm	m	LP	0.29
extra for butt weld	nr	LP	0.46
extra for square branch weld	nr	LP	0.66
extra for swept branch weld	nr	LP	1.20
extra for flange weld	nr	LP	0.69
50mm	m	LP	0.39
extra for butt weld	nr	LP	0.54
extra for square branch weld	nr	LP	0.77
extra for swept branch weld	nr	LP	1.48
extra for flange weld	nr	LP	0.90

Mild steel pipework, arc welding

Mild steel pipes with black heavy malleable iron fittings excluding supports, (flange welds include for two outer and one inner welds), nominal bore

	Unit	Labour grade	Labour hours
50mm	m	LP	0.39
extra for butt weld	nr	LP	0.25
extra for square branch weld	nr	LP	0.25
extra for swept branch weld	nr	LP	0.25
extra for flange weld	nr	LP	0.74
65mm	m	LP	0.47
extra for butt weld	nr	LP	0.65
extra for square branch weld	nr	LP	0.93
extra for swept branch weld	nr	LP	1.57
extra for flange weld	nr	LP	0.93

	Unit	Labour grade	Labour hours
80mm	m	LP	0.58
extra for butt weld	nr	LP	0.77
extra for square branch weld	nr	LP	1.36
extra for swept branch weld	nr	LP	2.09
extra for flange weld	nr	LP	1.05
100mm	m	LP	0.76
extra for butt weld	nr	LP	1.05
extra for square branch weld	nr	LP	1.73
extra for swept branch weld	nr	LP	2.28
extra for flange weld	nr	LP	1.39
125mm	m	LP	0.79
extra for butt weld	nr	LP	1.36
extra for square branch weld	nr	LP	2.04
extra for swept branch weld	nr	LP	2.62
extra for flange weld	nr	LP	1.73
150mm	m	LP	1.12
extra for butt weld	nr	LP	1.65
extra for square branch weld	nr	LP	2.50
extra for swept branch weld	nr	LP	3.09
extra for flange weld	nr	LP	2.07
200mm	m	LP	1.46
extra for butt weld	nr	LP	2.35
extra for square branch weld	nr	LP	3.70
extra for swept branch weld	nr	LP	4.32
extra for flange weld	nr	LP	2.70

	Unit	Labour grade	Labour hours

Pipework ancillaries

Cast iron gate valve with flanged ends, nominal bore

	Unit	Labour grade	Labour hours
50mm	nr	LP	0.77
65mm	nr	LP	0.93
80mm	nr	LP	1.05
100mm	nr	LP	1.15
125mm	nr	LP	1.30
150mm	nr	LP	1.40
200mm	nr	LP	4.00
250mm	nr	LP	6.00

Cast iron globe valve with flanged ends, nominal bore

	Unit	Labour grade	Labour hours
50mm	nr	LP	0.77
65mm	nr	LP	0.93
80mm	nr	LP	1.05
100mm	nr	LP	1.15
125mm	nr	LP	1.30
150mm	nr	LP	1.40
200mm	nr	LP	4.00
250mm	nr	LP	6.00

Cast iron flangeless butterfly valve, nominal bore

	Unit	Labour grade	Labour hours
50mm	nr	LP	0.77
65mm	nr	LP	0.93
80mm	nr	LP	1.05
100mm	nr	LP	1.15
125mm	nr	LP	1.30
150mm	nr	LP	1.40
200mm	nr	LP	4.00
250mm	nr	LP	6.00

	Unit	Labour grade	Labour hours
Cast iron double regulating valve, nominal bore			
50mm	nr	LP	0.77
65mm	nr	LP	0.93
80mm	nr	LP	1.05
100mm	nr	LP	1.15
125mm	nr	LP	1.30
150mm	nr	LP	1.40
200mm	nr	LP	4.00
250mm	nr	LP	6.00
Cast iron commissioning set, for pipe nominal bore			
50mm	nr	LP	0.77
65mm	nr	LP	0.93
80mm	nr	LP	1.05
100mm	nr	LP	1.15
125mm	nr	LP	1.30
150mm	nr	LP	1.40
200mm	nr	LP	4.00
250mm	nr	LP	6.00
Cast iron metering station, for pipe nominal bore			
50mm	nr	LP	0.77
65mm	nr	LP	0.93
80mm	nr	LP	1.05
100mm	nr	LP	1.15
125mm	nr	LP	1.30
150mm	nr	LP	1.40
200mm	nr	LP	4.00
250mm	nr	LP	6.00

	Unit	Labour grade	Labour hours
Stainless steel orifice plate for flanges, size			
50mm	nr	LP	0.77
65mm	nr	LP	0.93
80mm	nr	LP	1.05
100mm	nr	LP	1.15
125mm	nr	LP	1.30
150mm	nr	LP	1.40
200mm	nr	LP	4.00
250mm	nr	LP	6.00
Cast iron float-operated equilibrium ball valve with copper ball, for pipe nominal bore			
65mm	nr	LP	0.66
80mm	nr	LP	0.79
100mm	nr	LP	1.02
Cast iron swing pattern check valve with flanged ends, for pipe nominal			
50mm	nr	LP	0.77
65mm	nr	LP	0.93
80mm	nr	LP	1.05
100mm	nr	LP	1.15
125mm	nr	LP	1.30
150mm	nr	LP	1.40
200mm	nr	LP	4.00
250mm	nr	LP	6.00

	Unit	Labour grade	Labour hours
Bronze gate valve with female screwed			
20mm	nr	LP	0.49
25mm	nr	LP	0.56
32mm	nr	LP	0.62
40mm	nr	LP	0.71
50mm	nr	LP	0.77
Bronze gate valve with female ends, for pipe nominal bore			
50mm	nr	LP	0.77
65mm	nr	LP	0.93
80mm	nr	LP	1.05
100mm	nr	LP	1.15
Bronze oblique double regulating valve with female ends, for pipe nominal bore			
20mm	nr	LP	0.49
25mm	nr	LP	0.56
32mm	nr	LP	0.62
40mm	nr	LP	0.71
50mm	nr	LP	0.77
Bronze commissioning set, for pipe nominal bore			
20mm	nr	LP	0.49
25mm	nr	LP	0.56
32mm	nr	LP	0.62
40mm	nr	LP	0.71
50mm	nr	LP	0.77

	Unit	Labour grade	Labour hours
Bronze metering station with female screwed end, for pipe nominal bore			
20mm	nr	LP	0.49
25mm	nr	LP	0.56
32mm	nr	LP	0.62
40mm	nr	LP	0.71
50mm	nr	LP	0.77
Bronze angle patterned radiator valve			
15mm	nr	LP	0.46
20mm	nr	LP	0.49
25mm	nr	LP	0.56
Chromium-plated angle patterned radiator valve			
15mm	nr	LP	0.46
20mm	nr	LP	0.49
25mm	nr	LP	0.56
Gunmetal float-operated equilibrium ball valve with copper ball, for pipe nominal bore			
25mm	nr	LP	0.56
32mm	nr	LP	0.62
40mm	nr	LP	0.71
50mm	nr	LP	0.77
Bronze draincock with male screwed inlet			
15mm	nr	LP	0.46
20mm	nr	LP	0.49

	Unit	Labour grade	Labour hours
Bronze three-way gland cock with female screwed ends			
15mm	nr	LP	0.69
20mm	nr	LP	0.74
25mm	nr	LP	0.83
32mm	nr	LP	0.93
40mm	nr	LP	0.97
50mm	nr	LP	1.16
Stainless steel thermodynamic steam trap with in-built strainer and female screwed ends			
15mm	nr	LP	0.46
20mm	nr	LP	0.49
Gunmetal sight glass with double windows and female screwed ends			
15mm	nr	LP	0.46
20mm	nr	LP	0.49

	Unit	Labour grade	Labour hours

Expansion devices

Stainless steel axial type expansion
device with flanged ends fixed into
pipelines, nominal bore

	Unit	Labour grade	Labour hours
15mm	nr	LP	0.46
20mm	nr	LP	0.49
25mm	nr	LP	0.56
32mm	nr	LP	0.62
40mm	nr	LP	0.71
50mm	nr	LP	0.77
65mm	nr	LP	0.93
80mm	nr	LP	1.05
100mm	nr	LP	1.15
150mm	nr	LP	1.30
200mm	nr	LP	1.85

Synthetic rubber compensator with
reinforced flanged ends, tied, for pipes
nominal bore

	Unit	Labour grade	Labour hours
32mm	nr	LP	0.62
40mm	nr	LP	0.71
50mm	nr	LP	0.77
65mm	nr	LP	0.93
80mm	nr	LP	1.05
100mm	nr	LP	1.15
150mm	nr	LP	1.30

	Unit	Labour grade	Labour hours
Synthetic rubber compensator with reinforced flanged ends, untied, for pipes nominal bore			
32mm	nr	LP	0.62
40mm	nr	LP	0.71
50mm	nr	LP	0.77
65mm	nr	LP	0.93
80mm	nr	LP	1.05
100mm	nr	LP	1.15
150mm	nr	LP	1.30
Stainless steel synthetic flexible hose 300mm long, diameter			
9mm	nr	LP	0.40
15mm	nr	LP	0.46
20mm	nr	LP	0.49
Cast iron Y-pattern strainer with flanged ends, diameter			
50mm	nr	LP	0.77
65mm	nr	LP	0.93
80mm	nr	LP	1.05
100mm	nr	LP	1.15
125mm	nr	LP	1.30
150mm	nr	LP	1.48
200mm	nr	LP	1.85
250mm	nr	LP	2.32

	Unit	Labour grade	Labour hours

Gauges

	Unit	Labour grade	Labour hours
Bourdon tube-type pressure gauge with ring syphon and 100mm dial	nr	LP	0.40
Black steel dial pattern altitude gauge with 100mm dial	nr	LP	0.46

Pumps

Bronze in-line hot water service, 415 volt, three-phase glandless circulating pump

	Unit	Labour grade	Labour hours
25mm screwed, 0.8 l/s @ 27kPa	nr	LP	0.58
50mm flanged, 0.8 l/s @ 30kPa	nr	LP	1.39

Cast iron in-line LPHW, 415 volt, three-phase glandless circulating pump

	Unit	Labour grade	Labour hours
32mm screwed, 0.7 l/s @ 45kPa	nr	LP	0.58
50mm flanged, 2.5 l/s @ 65kPa	nr	LP	1.39

Cast iron in-line LPHW and chilled, 415 volt, three-phase glanded circulating pump

	Unit	Labour grade	Labour hours
25mm screwed, 0.5 l/s @ 15kPa	nr	LP	0.58
50mm flanged, 3.0 l/s @ 50kPa	nr	LP	1.39
80mm flanged, 7.0 l/s @ 100kPa	nr	LP	2.93
100mm flanged, 25.0 l/s @ 120kPa	nr	LP	3.47
150mm flanged, 70.0 l/s @ 85kPa	nr	LP	5.01

	Unit	Labour grade	Labour hours

Cast iron end suction belt-driven LPHW and chilled, 1450rpm motor, three phase 415 volt water pump fixed

	Unit	Labour grade	Labour hours
50/25mm flanged, 3.0 l/s @ 75kPa	nr	LP	2.70
50mm flanged, 6.0 l/s @ 85kPa	nr	LP	3.00
80mm flanged, 10.0 l/s @ 200kPa	nr	LP	4.00
100mm flanged, 25.0 l/s @ 220kPa	nr	LP	5.00
150mm flanged, 55.0 l/s @ 400kPa	nr	LP	6.00
200mm flanged, 100.0 l/s @ 200kPa	nr	LP	9.00
Extra for fixings	nr	LP	1.88

Cast iron horizontal end close-coupled LPHW and chilled, slide rail 1450 rpm motor, three phase 415 volt water pump fixed to metal

50/25mm flanged, 3.0 l/s @ 60kPa	nr	LP	2.70
50mm flanged, 8.0 l/s @ 140kPa	nr	LP	3.00
80mm flanged, 12.0 l/s @ 250kPa	nr	LP	4.00
100mm flanged, 25.0 l/s @ 200kPa	nr	LP	5.00
Extra for fixings	nr	LP	1.88

Cast iron twin in-line glandless standby LPHW, three phase 415 volt pump fixed to masonry

32mm screwed, 0.8 l/s @ 30kPa	nr	LP	2.70
50mm flanged, 2.0 l/s @ 90kPa	nr	LP	3.00
80mm flanged, 16.0 l/s @ 70kPa	nr	LP	4.00

	Unit	Labour grade	Labour hours
Cast iron dual in-line LPHW and chilled, manifolds, valves, three phase 415 volt pump fixed to masonry			
50mm flanged, 25.0 l/s @ 30kPa	nr	LP	2.70
50mm flanged, 4.0 l/s @ 45kPa	nr	LP	3.00
80mm flanged, 8.0 l/s @ 80kPa	nr	LP	4.00
100mm flanged, 25.0 l/s @ 90kPa	nr	LP	5.00
Extra for fixings	nr	LP	1.88
Cast iron duty/standby booster set, multi-stage cold water pumps, valves, diaphragm tank, panel, valves, three phase 415 volt pump fixed to masonry			
80mm flanged, 1.0 l/s @ 600kPa	nr	LP	4.50
80mm flanged, 3.5 l/s @ 200kPa	nr	LP	4.50
150mm flanged, 25.0 l/s @ 1000kPa	nr	LP	4.50
Extra for fixings	nr	LP	1.88
Cast iron duty/standby booster set, end suction, cold water pumps, valves, diaphragm tank, panel, valves, three phase 415 volt pump fixed to masonry			
80mm flanged, 4.0 l/s @ 600kPa	nr	LP	4.50
100mm flanged, 10.5 l/s @ 550kPa	nr	LP	5.50
80mm flanged, 1.0 l/s @ 1000kPa	nr	LP	6.50
Extra for fixings	nr	LP	1.88

	Unit	Labour grade	Labour hours
Bronze sealed packaged automatic LPHW make-up unit, 415 volt, three phase, fixed to masonry			
15mm screwed, 700 litres with			
10 metres static head	nr	LP	4.00
15mm screwed, 2000 litres with			
10 metres static head	nr	LP	5.00
Extra for fixings	nr	LP	1.88
Bronze sealed packaged pressurisation unit, chilled water, automatic make-up unit, 415 volt, three phase, fixed to			
15mm screwed, 2000 litres with			
10 metres static head	nr	LP	5.00
15mm screwed, 10,000 litres with			
10 metres static head	nr	LP	6.00
Extra for fixings	nr	LP	1.88
Cast iron packaged fire hose pumping set, duty/standby pumps, valves, diaphragm tank, control pane, 415 volt, three phase, fixed to masonry			
80mm flanged, 2.3 l/s @ 290kPa	nr	LP	4.00
80mm flanged, 2.3 l/s @ 720kPa	nr	LP	5.00
Extra for fixings	nr	LP	1.88
Stainless steel submersible sump pump, single phase, placed in position			
12mm screwed, 2.5 l/s	nr	LP	0.83

	Unit	Labour grade	Labour hours

Tanks, cisterns and cylinders

Heavy duty plastic oil storage tank with manhole cover and connections to services

	Unit	Labour grade	Labour hours
2100 x 1000 x 1000mm, 1360 litres	nr	LP	7.72
2150 x 1500 x 1300mm, 2460 litres	nr	LP	7.72

Galvanised mild steel oil storage tank with manhole cover and connections to services finished with one coat bituminous paint

1800 x 600 x 1200mm, 1360 litres	nr	LP	7.72
1800 x 1200 x 1200mm, 2730 litres	nr	LP	7.72
2450 x 1500 x 1200mm, 4550 litres	nr	LP	12.35

Galvanised mild steel tank with open top and loose cover

18 litres ref. SCM 45	nr	LP	1.40
36 litres ref. SCM 70	nr	LP	1.40
54 litres ref. SCM 90	nr	LP	1.40
68 litres ref. SCM 110	nr	LP	1.40
86 litres ref. SCM 135	nr	LP	1.40
114 litres ref. SCM 180	nr	LP	1.40
159 litres ref. SCM 230	nr	LP	1.40
191 litres ref. SCM 270	nr	LP	1.40
227 litres ref. SCM 320	nr	LP	1.70
264 litres ref. SCM 360	nr	LP	1.70
327 litres ref. SCM 450	nr	LP	1.70
423 litres ref. SCM 570	nr	LP	1.70
491 litres ref. SCM 680	nr	LP	2.78

	Unit	Labour grade	Labour hours
709 litres ref. SCM 910	nr	LP	2.78
841 litres ref. SCM 1130	nr	LP	3.47
1227 litres ref. SCM 1600	nr	LP	3.47
1727 litres ref. SCM 2270	nr	LP	4.63
2137 litres ref. SCM 2720	nr	LP	4.63

Copper hot water insulated direct cylinder, grade 3

	Unit	Labour grade	Labour hours
1050mm high x 400mm diameter, capacity 115 litres	nr	LP	1.93
900mm high x 400mm diameter, capacity 120 litres	nr	LP	1.93

Copper hot water insulated direct cylinder, grade 4

	Unit	Labour grade	Labour hours
1050mm high x 350mm diameter, capacity 88 litres	nr	LP	2.70
900mm high x 400mm diameter, capacity 96 litres	nr	LP	2.70
1050mm high x 400mm diameter, capacity 114 litres	nr	LP	2.70
900mm high x 450mm diameter, capacity 117 litres	nr	LP	2.70
1050mm high x 450mm diameter, capacity 140 litres	nr	LP	2.70
1200mm high x 450mm diameter, capacity 190 litres	nr	LP	2.70
1500mm high x 500mm diameter, capacity 245 litres	nr	LP	3.00
1200mm high x 600mm diameter, capacity 280 litres	nr	LP	3.00
1500mm high x 600mm diameter, capacity 360 litres	nr	LP	3.00

	Unit	Labour grade	Labour hours
Copper hot water insulated indirect combination cylinder			
900mm high x 450mm diameter, capacity 85 litres	nr	LP	2.70
1050mm high x 450mm diameter, capacity 115 litres	nr	LP	2.70
1200mm high x 450mm diameter, capacity 115 litres	nr	LP	2.70
1400mm high x 500mm diameter, capacity 115 litres	nr	LP	2.70
Mild steel vertical pattern copper-lined calorifier, primary LPHW, 10 to 65 degrees in one hour, 3 bar primary maximum working pressure, capacity			
2000 litres	nr	LP	6.55
3000 litres	nr	LP	9.72
4000 litres	nr	LP	13.50
5000 litres	nr	LP	14.28
6000 litres	nr	LP	15.00
7000 litres	nr	LP	16.98
Copper vertical pattern calorifier, primary LPHW, 10 to 65 degrees in one hour, 3 bar primary maximum working pressure, capacity			
2000 litres	nr	LP	6.55
3000 litres	nr	LP	9.72
4000 litres	nr	LP	13.50
5000 litres	nr	LP	14.28
6000 litres	nr	LP	15.00
7000 litres	nr	LP	16.98

	Unit	Labour grade	Labour hours
Immersion heater, rod type, 2.25BSP head thermostat			
1 kW, 280mm long	nr	LP	0.67
2 kW, 380mm long	nr	LP	0.67
3 kW, 280mm long	nr	LP	0.67
4 kW, 405mm long	nr	LP	0.80
6 kW, 760mm long	nr	LP	0.96
6 kW, 1065mm long	nr	LP	0.96
7.5 kW, 610mm long	nr	LP	1.16
9 kW, 915mm long	nr	LP	1.39
12 kW, 610mm long	nr	LP	1.67
18 kW, 915mm long	nr	LP	2.00

Boilers

Domestic central heating boiler, wall-mounted, electric controls, conventional flue, natural gas-fired			
8.78Kw/h	nr	LP	5.40
11.70Kwh	nr	LP	5.40
14.62Kw/h	nr	LP	5.40
17.55Kw/h	nr	LP	5.40
Extra for flue	m	LP	0.70
Domestic central heating boiler, wall-mounted, electric controls, fan-assisted flue, natural gas-fired			
8.78Kw/h	nr	LP	5.40
11.70Kw/h	nr	LP	5.40
14.62Kw/h	nr	LP	5.40
17.55Kw/h	nr	LP	5.40

	Unit	Labour grade	Labour hours
Domestic central heating boiler, floor-standing, electric controls, conventional flue, natural gas-fired			
11.70Kw/h	nr	LP	4.63
14.62Kw/h	nr	LP	4.63
20.48Kw/h	nr	LP	4.63
23.40Kw/h	nr	LP	5.01
29.25Kw/h	nr	LP	4.50
36.56Kw/h	nr	LP	6.17
Extra for flue	m	LP	6.17
Domestic central heating boiler, floor-standing, electric controls, balanced flue, natural gas-fired			
11.70Kw/h	nr	LP	5.40
14.62Kw/h	nr	LP	5.40
20.48Kw/h	nr	LP	5.40
23.40Kw/h	nr	LP	6.17
Industrial natural gas-fired hot water boiler with enamelled casing, controls, mountings, burners and insulation			
147kW	nr	LP	7.72
175kW	nr	LP	7.72
200kW	nr	LP	7.72
230kW	nr	LP	7.72
270kW	nr	LP	7.72
293kW	nr	LP	9.26
360kW	nr	LP	9.26
500kW	nr	LP	9.26
600kW	nr	LP	12.34
750kW	nr	LP	18.52
1000kW	nr	LP	21.60

	Unit	Labour grade	Labour hours
1150kW	nr	LP	23.15
1500kW	nr	LP	24.70
1760kW	nr	LP	26.24
2050kW	nr	LP	29.33
2350kW	nr	LP	32.40
3000kW	nr	LP	35.50

Industrial oil-fired hot water boiler with enamelled casing, controls, mountings, burners and insulation

	Unit	Labour grade	Labour hours
147kW	nr	LP	7.72
175kW	nr	LP	7.72
200kW	nr	LP	7.72
230kW	nr	LP	7.72
270kW	nr	LP	7.72
293kW	nr	LP	9.26
360kW	nr	LP	9.26
500kW	nr	LP	9.26
600kW	nr	LP	12.34
750kW	nr	LP	18.52
1000kW	nr	LP	21.60
1150kW	nr	LP	23.15
1500kW	nr	LP	24.70
1760kW	nr	LP	26.24
2050kW	nr	LP	29.33
2350kW	nr	LP	32.40
3000kW	nr	LP	35.50

	Unit	Labour grade	Labour hours

Radiators

Single panel steel radiator,15mm thick,
fixed to brackets screwed to walls

	Unit	Labour grade	Labour hours
320mm high, 500mm long, 226 watts	nr	LP	0.77
320mm high, 750mm long, 339 watts	nr	LP	0.77
320mm high, 1000mm long, 452 watts	nr	LP	0.77
320mm high, 1500mm long, 678 watts	nr	LP	0.77
320mm high, 2000mm long, 904 watts	nr	LP	1.16
320mm high, 2500mm long, 1130 watts	nr	LP	1.16
420mm high, 500mm long, 291 watts	nr	LP	1.16
420mm high, 750mm long, 437 watts	nr	LP	1.16
420mm high, 1000mm long, 582 watts	nr	LP	1.16
420mm high, 1500mm long, 873 watts	nr	LP	1.16
420mm high, 2000mm long, 1164 watts	nr	LP	1.59
420mm high, 2500mm long, 1455 watts	nr	LP	1.59
520mm high, 500mm long, 354 watts	nr	LP	1.59
520mm high, 750mm long, 531 watts	nr	LP	1.59
520mm high, 1000mm long, 708 watts	nr	LP	1.59
520mm high, 1500mm long, 1062 watts	nr	LP	1.59
520mm high, 2000mm long, 1416 watts	nr	LP	1.93
520mm high, 2500mm long, 1770 watts	nr	LP	1.93

Double panel steel radiator, 15mm
thick, fixed to brackets screwed to walls

	Unit	Labour grade	Labour hours
320mm high, 500mm long, 405 watts	nr	LP	1.16
320mm high, 750mm long, 608 watts	nr	LP	1.16
320mm high, 1000mm long, 810 watts	nr	LP	1.16
320mm high, 1500mm long, 1215 watts	nr	LP	1.16
320mm high, 2000mm long, 1620 watts	nr	LP	1.54
320mm high, 2500mm long, 2025 watts	nr	LP	1.54

	Unit	Labour grade	Labour hours
420mm high, 500mm long, 515 watts	nr	LP	1.54
420mm high, 750mm long, 772 watts	nr	LP	1.54
420mm high, 1000mm long, 1029 watts	nr	LP	1.54
420mm high, 1500mm long, 1544 watts	nr	LP	1.54
420mm high, 2000mm long, 2058 watts	nr	LP	1.93
420mm high, 2500mm long, 2573 watts	nr	LP	1.93
520mm high, 500mm long, 620 watts	nr	LP	1.93
520mm high, 750mm long, 930 watts	nr	LP	1.93
520mm high, 1000mm long, 1240 watts	nr	LP	1.93
520mm high, 1500mm long, 1860 watts	nr	LP	1.93
520mm high, 2000mm long, 2480 watts	nr	LP	2.32
520mm high, 2500mm long, 3100 watts	nr	LP	2.32

Refrigeration units

Air conditioning unit, with cassette and fan coil unit, remote fitted rotary-type compressor unit, cooling only, based on outdoor temperature 28 degrees dry

	Unit	Labour grade	Labour hours
3.3kW cooling	nr	LP	4.00
4.9kW cooling	nr	LP	4.50
6.2kW cooling	nr	LP	5.00
7.2kW cooling	nr	LP	5.50
9.9kW cooling	nr	LP	6.00
12.2kW cooling	nr	LP	6.50

	Unit	Labour grade	Labour hours
Air conditioning unit, with cassette and fan coil unit, remote fitted rotary-type compressor unit, cooling and heat pump, based on outdoor temperature 28 degrees dry bulb			
3.6kW cooling, 3.8kW heating	nr	LP	4.00
4.7kW cooling, 4.8kW heating	nr	LP	4.50
6.6kW cooling, 6.6kW heating	nr	LP	5.00
7.2kW cooling, 7.9kW heating	nr	LP	5.50
9.8kW cooling, 11.2kW heating	nr	LP	6.00
12.2kW cooling, 14.0kW heating	nr	LP	6.50
Air conditioning unit, wall-mounted with fan coil unit, split system direct expansion, remote fitted rotary-type compressor unit, cooling only, based on outdoor temperature 28 degrees dry bulb			
1.8kW cooling	nr	LP	3.50
2.3kW cooling	nr	LP	4.00
3.3kW cooling	nr	LP	4.50
4.9kW cooling	nr	LP	5.00
6.2kW cooling	nr	LP	5.50
7.2kW cooling	nr	LP	6.00
9.8kW cooling	nr	LP	6.50

	Unit	Labour grade	Labour hours

Air conditioning unit, wall-mounted with fan coil unit, split system direct expansion, remote fitted rotary-type compressor unit, cooling and heat pump, based on outdoor temperature 28 degrees dry bulb

	Unit	Labour grade	Labour hours
2.4kW cooling, 2.8kW heating	nr	LP	4.00
3.6kW cooling, 3.6kW heating	nr	LP	4.50
4.7kW cooling, 4.8kW heating	nr	LP	5.00
6.1kW cooling, 7.7kW heating	nr	LP	5.50
7.2kW cooling, 7.9kW heating	nr	LP	6.00
9.8kW cooling, 11.2kW heating	nr	LP	6.50

Air-cooled reciprocating compressor, multi-step by cylinder unloading or twin-speed compressor, shell and tube evaporator, control starter panel, 30C degrees summer and minus 4C degrees winter ambient temperatures, chilled water flow 6-12C degrees

	Unit	Labour grade	Labour hours
25kW, 7.0 tons	nr	LP	4.00
50kW, 14.0 tons	nr	LP	4.50
75kW, 21.0 tons	nr	LP	5.00
100kW, 28.5 tons	nr	LP	5.50
125kW, 35.5 tons	nr	LP	6.00
150kW, 42.7 tons	nr	LP	6.50
175kW, 49.8 tons	nr	LP	7.00
200kW, 56.9 tons	nr	LP	8.00
225kW, 64.0 tons	nr	LP	9.00
250kW, 71.0 tons	nr	LP	10.00
300kW, 85.0 tons	nr	LP	11.00
400kW, 114.0 tons	nr	LP	12.00
500kW, 142.0 tons	nr	LP	13.00

	Unit	Labour grade	Labour hours
600kW, 170.0 tons	nr	LP	14.00
750kW, 214.0 tons	nr	LP	15.00
1000kW, 285.0 tons	nr	LP	16.00

Water-cooled centrifugal single
compressor, force-feed lubrication
system, shell and tube-type condensor,
evaporator, flooded-type heat
exchanger, chilled water flow 2-12C
degrees, condenser flow 27-32C
degrees

	Unit	Labour grade	Labour hours
200kW, 56.9 tons	nr	LP	10.00
300kW, 85.3 tons	nr	LP	12.50
400kW, 113.7 tons	nr	LP	15.00
500kW, 142.2 tons	nr	LP	17.50
600kW, 170.6 tons	nr	LP	20.00
700kW, 199.9 tons	nr	LP	22.50
800kW, 227.5 tons	nr	LP	25.00
900kW, 255.9 tons	nr	LP	27.50
1000kW, 284.5 tons	nr	LP	30.00
1500kW, 426.0 tons	nr	LP	35.00

Water-cooled reciprocating compressor,
multi-step by cylinder unloading or
twin-speed compressor, shell and tube
evaporator, control starter panel, 30C
degrees summer and minus 4C degrees
winter ambient temperatures, chilled
water flow 6-12C degrees

	Unit	Labour grade	Labour hours
25kW, 7.0 tons	nr	LP	5.00
50kW, 14.0 tons	nr	LP	5.50
75kW, 21.0 tons	nr	LP	6.00
100kW, 28.5 tons	nr	LP	6.50

	Unit	Labour grade	Labour hours
125kW, 35.5 tons	nr	LP	7.00
150kW, 42.7 tons	nr	LP	7.50
175kW, 49.8 tons	nr	LP	8.00
200kW, 56.9 tons	nr	LP	8.50
225kW, 64.0 tons	nr	LP	9.00
250kW, 71.0 tons	nr	LP	9.50

Cooling towers, forced draught
counterflow type, belt-driven
centrifugal fan, flexible connections,
anti-vibration mountings, coolings from
32C to 27C degrees, ambient air wet
bulb 20C degrees

	Unit	Labour grade	Labour hours
100kW	nr	LP	10.00
200kW	nr	LP	12.50
300kW	nr	LP	15.00
400kW	nr	LP	17.50
500kW	nr	LP	20.00
600kW	nr	LP	21.50
700kW	nr	LP	23.00
800kW	nr	LP	24.50
900kW	nr	LP	26.00
1000kW	nr	LP	27.50

38

Ventilation systems

Labour gangs

1 Foreman, 1 advance fitter/welder
(gas/arc), 2 advanced fitters (gas or arc),
3 advanced fitters, 2 fitter and   1
mate   LP

	Unit	Labour grade	Labour hours

Fans

Axial flow fan, single stage, 415 volt three phase, galvanised finish, fixed in ductlines

0.5m3/s @ 200Pa	nr	LP	7.40
1.0m3/s @ 200Pa	nr	LP	6.87
2.5m3/s @ 200Pa	nr	LP	8.03
3.0m3/s @ 200Pa	nr	LP	9.41
4.0m3/s @ 200Pa	nr	LP	10.80

Axial flow fan, single stage, 415 volt thrtee phase, galvanised finish, fixed in ductlines

0.5m3/s @ 200Pa	nr	LP	7.40
1.0m3/s @ 200Pa	nr	LP	6.87
2.5m3/s @ 200Pa	nr	LP	8.03
3.0m3/s @ 200Pa	nr	LP	9.41
4.0m3/s @ 200Pa	nr	LP	10.80

Centrifugal fan, backward curved, 415 volt three phase, wedge belt drive, galvanised finish, fixed in ductlines

0.5m3/s @ 200Pa	nr	LP	12.80
1.0m3/s @ 200Pa	nr	LP	13.74
2.5m3/s @ 200Pa	nr	LP	15.44
3.0m3/s @ 200Pa	nr	LP	16.78
4.0m3/s @ 200Pa	nr	LP	21.76

	Unit	Labour grade	Labour hours
5.0m3/s @ 200Pa	nr	LP	23.00
7.5m3/s @ 200Pa	nr	LP	26.70
10.0m3/s @ 200Pa	nr	LP	29.64
12.5m3/s @ 200Pa	nr	LP	33.03
15.0m3/s @ 200Pa	nr	LP	41.98

Roof extract unit mixed flow fan, 415
volt three phase direct drive, vertical
discharge fixed to background

300mm diameter, 0.5m3/s @ 200Pa	nr	LP	7.25
300mm diameter, 1.0m3/s @ 200Pa	nr	LP	7.25
400mm diameter, 2.5m3/s @ 200Pa	nr	LP	7.70
500mm diameter, 3.0m3/s @ 200Pa	nr	LP	8.60
750mm diameter, 4.0m3/s @ 200Pa	nr	LP	11.58

Smoke extract unit, axial flow fan, 415
volt three phase, galvanised finish,
fixed in ductlines

0.5m3/s @ 200Pa	nr	LP	7.40
1.0m3/s @ 200Pa	nr	LP	6.87
2.5m3/s @ 200Pa	nr	LP	8.03
3.0m3/s @ 200Pa	nr	LP	9.41
4.0m3/s @ 200Pa	nr	LP	10.80

	Unit	Labour grade	Labour hours

Twin extract unit, 415 volt three phase, belt drive, automatic changeover, galvanised finish, fixed in ductlines

	Unit	Labour grade	Labour hours
0.5m3/s @ 200Pa	nr	LP	1.54
1.0m3/s @ 200Pa	nr	LP	1.85
2.5m3/s @ 200Pa	nr	LP	3.09
3.0m3/s @ 200Pa	nr	LP	3.86
4.0m3/s @ 200Pa	nr	LP	4.63

Starters, three phase, direct on-line, manually operated, fractional horse-power

	Unit	Labour grade	Labour hours
1 to 3	nr	LP	1.55
3 to 10	nr	LP	1.98
11 to 20	nr	LP	2.42
21 to 30	nr	LP	2.96
31 to 40	nr	LP	3.05
41 to 50	nr	LP	3.50
51 to 100	nr	LP	4.00
101 to 300	nr	LP	4.85

Domestic fans, fixed in windows, back-draught shutter, remote controller, 240 volt single phase, diameter

	Unit	Labour grade	Labour hours
150mm	nr	LP	1.17
230mm	nr	LP	1.17

	Unit	Labour grade	Labour hours
Domestic fans, wall-mounted, back-draught shutter, remote controller, 240 volt single phase, diameter			
150mm	nr	LP	1.33
230mm	nr	LP	1.33

PART FIVE

ELECTRICAL WORK

39

General electrical information

Lighting

The formula for assessing the amount of illumunation required is usually described as the lumen method and is expressed as:

$$F \quad = \quad \frac{A \times Eav}{CU \times M}$$

where
F	=	is the total number of lumens required
A	=	is the area to be illuminated
Eav	=	is the average illumination on the working plane
CU	=	is the coefficient of illumination and
M	=	is the maintenance factor.

The maintenance factor is usually stated as 0.8 but can be 0.6 in dirty areas. The figures to be applied to Eav are based upon tables of average illumination levels in different working conditions.

Area	Lux	Lumens/ft2
General office conditions	500	50
Drawing office	750	75
Corridoors, store rooms	300	30
Shop counters	500	50
Watch repairing	3000	300
Proof reading	750	75
Living rooms	100	10
Bedrooms	50	5

To select the coefficient of utilisation (CU) it is necessary to determine the room index.

$$RI \quad = \quad \frac{L \times W}{Hm(L \times W)}$$

where   RI   =   is the room index
             L   =   is the length of the room
             W   =   is the width of the room and
             Hm   =   is the height of the fitting above the working plane

A working height of 0.85m above floor level should be allowed where the working plane is a desk or bench top. The coefficient of utilisation (CU) is selected from the manufacturer's design information and after calculating the total illumination required, the design lumens can be chosen and the number of light fittings can be assessed.

$$\text{Number of light fittings} \quad = \quad \frac{\underline{\text{Illumination required (F)}}}{\text{Design lumens per lamp}}$$

Cable capacity

Number of cables	Conductor area mm2	Cable diameter mm	Total cross-sectional area mm2
1	1.0	2.8	6.2
2	1.0	2.8	12.3
3	1.0	2.8	18.5
4	1.0	2.8	24.6
5	1.0	2.8	30.8
6	1.0	2.8	36.9
7	1.0	2.8	43.1
8	1.0	2.8	49.2
9	1.0	2.8	55.4
10	1.0	2.8	61.5
1	1.5	3.5	9.6
2	1.5	3.5	19.2
3	1.5	3.5	28.8
4	1.5	3.5	38.5
5	1.5	3.5	48.1
6	1.5	3.5	57.7
7	1.5	3.5	67.3
8	1.5	3.5	76.9
9	1.5	3.5	86.5
10	1.5	3.5	96.2

Number of cables	Conductor area mm2	Cable diameter mm	Total cross-sectional area mm2
1	2.5	4.2	13.8
2	2.5	4.2	27.7
3	2.5	4.2	41.5
4	2.5	4.2	55.4
5	2.5	4.2	69.2
6	2.5	4.2	83.1
7	2.5	4.2	96.9
8	2.5	4.2	110.8
9	2.5	4.2	124.6
10	2.5	4.2	138.5
1	4.0	4.8	18.1
2	4.0	4.8	36.2
3	4.0	4.8	54.3
4	4.0	4.8	72.3
5	4.0	4.8	90.4
6	4.0	4.8	108.5
7	4.0	4.8	126.6
8	4.0	4.8	144.7
9	4.0	4.8	162.8
10	4.0	4.8	180.9
1	6.0	5.4	22.9
2	6.0	5.4	45.8
3	6.0	5.4	68.7
4	6.0	5.4	91.6
5	6.0	5.4	114.5
6	6.0	5.4	137.3
7	6.0	5.4	160.2
8	6.0	5.4	183.1
9	6.0	5.4	206.0
10	6.0	5.4	363.0

Number of cables	Conductor area mm2	Cable diameter mm	Total cross-sectional area mm2
1	10.0	6.8	36.3
2	10.0	6.8	72.6
3	10.0	6.8	108.9
4	10.0	6.8	145.2
5	10.0	6.8	181.5
6	10.0	6.8	217.8
7	10.0	6.8	254.1
8	10.0	6.8	290.4
9	10.0	6.8	326.7
10	10.0	6.8	363.0

Labour grades/notional team

1 Technician, 1 approved electrician,
1 electrician, 1 apprentice
(18 year old) and 1 unskilled
operative   LQ

40

Conduits, cable trunking and trays

	Unit	Labour grade	Labour hours
Conduits			
Steel conduit, 20mm diameter, heavy gauge, complete with all necessary boxes, lids, bushes and the like, all measured in the running length			
Run on surface	m	LQ	0.65
Run in wall chase or floor screed	m	LQ	0.50
Run in floor slab	m	LQ	0.50
Steel conduit, 25mm diameter, heavy gauge, complete with all necessary boxes, lids, bushes and the like, all measured in the running length			
Run on surface	m	LQ	0.75
Run in wall chase or floor screed	m	LQ	0.60
Run in floor slab	m	LQ	0.60

	Unit	Labour grade	Labour hours

Steel conduit, 32mm diameter, heavy gauge, complete with all necessary boxes, lids, bushes and the like, all measured in the running length

Run on surface	m	LQ	1.00
Run in wall chase or floor screed	m	LQ	0.70
Run in floor slab	m	LQ	0.70

PVC conduit, 16mm diameter, complete with all necessary boxes, lids, bushes, adhesives and the like, all measured in the running length

Run on surface	m	LQ	0.60
Run in wall chase or floor screed	m	LQ	0.45
Run in floor slab	m	LQ	0.45

PVC conduit, 20mm diameter, complete with all necessary boxes, lids, bushes, adhesives and the like, all measured in the running length

Run on surface	m	LQ	0.60
Run in wall chase or floor screed	m	LQ	0.45
Run in floor slab	m	LQ	0.45

PVC conduit, 25mm diameter, complete with all necessary boxes, lids, bushes, adhesives and the like, all measured in the running length

Run on surface	m	LQ	0.60
Run in wall chase or floor screed	m	LQ	0.45
Run in floor slab	m	LQ	0.45

	Unit	Labour grade	Labour hours

PVC conduit, 32mm diameter, complete
with all necessary boxes, lids, bushes,
adhesives and the like, all measured in
the running length

	Unit	Labour grade	Labour hours
Run on surface	m	LQ	0.95
Run in wall chase or floor screed	m	LQ	0.65
Run in floor slab	m	LQ	0.60

Flexible metallic PVC-covered conduit,
complete with all necessary boxes, lids,
bushes, termination boxes and the like,
all measured in the running length, run
on surface, diameter

	Unit	Labour grade	Labour hours
16mm	m	LQ	1.00
20mm	m	LQ	1.00
25mm	m	LQ	1.50
32mm	m	LQ	2.00
38mm	m	LQ	2.50
50mm	m	LQ	3.00

Flexible metallic PVC-covered conduit,
complete with all necessary boxes, lids,
bushes, termination boxes and the like,
all measured in the running length, run
on surface, diameter

	Unit	Labour grade	Labour hours
16mm	m	LQ	1.00
20mm	m	LQ	1.00
25mm	m	LQ	1.50
32mm	m	LQ	2.00
38mm	m	LQ	2.50
50mm	m	LQ	3.00

	Unit	Labour grade	Labour hours

Adaptable boxes, fixed to backgrounds, size

	Unit	Labour grade	Labour hours
50 x 50 x 50mm deep	nr	LQ	0.30
75 x 75 x 50mm deep	nr	LQ	0.40
100 x 75 x 50mm deep	nr	LQ	0.50
100 x 100 x 75mm deep	nr	LQ	0.50
150 x 150 x 50mm deep	nr	LQ	0.65
150 x 150 x 75mm deep	nr	LQ	0.65
225 x 225 x 50mm deep	nr	LQ	0.80
225 x 225 x 100mm deep	nr	LQ	0.80
225 x 225 x 150mm deep	nr	LQ	0.85
300 x 300 x 100mm deep	nr	LQ	1.00
300 x 300 x 150mm deep	nr	LQ	1.10

Cable trunking

Galvanised steel single compartment trunking, including cover plates, earth continuity straps, supports, bends, tees, stop ends and the like, measured in the running length, fixing to background

	Unit	Labour grade	Labour hours
50 x 50mm	m	LQ	0.41
75 x 75mm	m	LQ	0.50
100 x 75mm	m	LQ	0.55
100 x 100mm	m	LQ	0.60
150 x 100mm	m	LQ	0.70
150 x 150mm	m	LQ	0.77

	Unit	Labour grade	Labour hours

Galvanised steel twin compartment trunking, including loose fillet plate, cover plates, earth continuity straps, supports, bends, tees, stop ends and the like, measured in the running length, fixing to background

50 x 50mm	m	LQ	0.46
75 x 75mm	m	LQ	0.55
100 x 75mm	m	LQ	0.60
100 x 100mm	m	LQ	0.65
150 x 100mm	m	LQ	0.75
150 x 150mm	m	LQ	0.82

Galvanised steel triple compartment trunking, including cover plates, earth continuity straps, supports, bends, tees, stop ends and the like, measured in the running length, fixing to background

50 x 50mm	m	LQ	0.51
75 x 75mm	m	LQ	0.60
100 x 75mm	m	LQ	0.65
100 x 100mm	m	LQ	0.70
150 x 100mm	m	LQ	0.80
150 x 150mm	m	LQ	0.87

White heavy duty PVC triple compartment bench trunking, including bends, tees, stop ends and the like, measured in the running length, fixing to background

212 x 50mm	m	LQ	1.24

	Unit	Labour grade	Labour hours
White heavy duty PVC triple compartment skirting trunking, including supports, bends, tees, stop ends and the like, measured in the running length, fixing to background			
212 x 50mm	m	LQ	1.29

Cable trays

Hot dip galvanised light/medium straight cable tray fixed to prepared supports measured separately, width			
50mm width, 1.0mm gauge	m	LQ	0.25
50mm width, 1.5mm gauge	m	LQ	0.25
75mm width, 1.0mm gauge	m	LQ	0.28
75mm width, 1.5mm gauge	m	LQ	0.28
100mm width, 1.0mm gauge	m	LQ	0.30
100mm width, 1.5mm gauge	m	LQ	0.30
150mm width, 1.0mm gauge	m	LQ	0.34
150mm width, 1.5mm gauge	m	LQ	0.34
225mm width, 1.2mm gauge	m	LQ	0.40
225mm width, 1.5mm gauge	m	LQ	0.40
300mm width, 1.5mm gauge	m	LQ	0.46
450mm width, 1.5mm gauge	m	LQ	0.46
450mm width, 2.0mm gauge	m	LQ	0.57
600mm width, 2.0mm gauge	m	LQ	0.70
750mm width, 2.0mm gauge	m	LQ	0.80
900mm width, 2.0mm gauge	m	LQ	0.92

	Unit	Labour grade	Labour hours
Flat bends			
50mm width, 1.5mm gauge	m	LQ	0.25
75mm width, 1.5mm gauge	m	LQ	0.28
100mm width, 1.5mm gauge	m	LQ	0.28
150mm width, 1.5mm gauge	m	LQ	0.30
225mm width, 1.5mm gauge	m	LQ	0.32
300mm width, 1.5mm gauge	m	LQ	0.36
450mm width, 2.0mm gauge	m	LQ	0.43
600mm width, 2.0mm gauge	m	LQ	0.50
750mm width, 2.0mm gauge	m	LQ	0.60
900mm width, 2.0mm gauge	m	LQ	0.75
Risers			
50mm width, 1.5mm gauge	m	LQ	0.38
75mm width, 1.5mm gauge	m	LQ	0.40
100mm width, 1.5mm gauge	m	LQ	0.48
150mm width, 1.5mm gauge	m	LQ	0.51
225mm width, 1.5mm gauge	m	LQ	0.60
300mm width, 1.5mm gauge	m	LQ	0.69
450mm width, 2.0mm gauge	m	LQ	0.86
600mm width, 2.0mm gauge	m	LQ	1.05
750mm width, 2.0mm gauge	m	LQ	1.25
900mm width, 2.0mm gauge	m	LQ	1.38

	Unit	Labour grade	Labour hours
Tees			
50mm width, 1.5mm gauge	m	LQ	0.25
75mm width, 1.5mm gauge	m	LQ	0.25
100mm width, 1.5mm gauge	m	LQ	0.28
150mm width, 1.5mm gauge	m	LQ	0.30
225mm width, 1.5mm gauge	m	LQ	0.32
300mm width, 1.5mm gauge	m	LQ	0.36
450mm width, 2.0mm gauge	m	LQ	0.43
600mm width, 2.0mm gauge	m	LQ	0.50
750mm width, 2.0mm gauge	m	LQ	0.75
900mm width, 2.0mm gauge	m	LQ	1.05
Four-way cross pieces			
50mm width, 1.5mm gauge	m	LQ	0.31
75mm width, 1.5mm gauge	m	LQ	0.31
100mm width, 1.5mm gauge	m	LQ	0.35
150mm width, 1.5mm gauge	m	LQ	0.38
225mm width, 1.5mm gauge	m	LQ	0.40
300mm width, 1.5mm gauge	m	LQ	0.45
450mm width, 2.0mm gauge	m	LQ	0.54
600mm width, 2.0mm gauge	m	LQ	0.63
750mm width, 2.0mm gauge	m	LQ	0.94
900mm width, 2.0mm gauge	m	LQ	1.31

	Unit	Labour grade	Labour hours

Fixings, brackets and supports

Unistrut (P1000), cut into pieces up to 1 metre long	nr	LQ	0.60
Unistrut (P1000), cut into pieces up to 1 metre long, to backgrounds requiring fixings (2 nr)	nr	LQ	1.07

Fixings only

Self-drill anchors, diameter

10mm	nr	LQ	0.15
12mm	nr	LQ	0.17
16mm	nr	LQ	0.21
20mm	nr	LQ	0.26

Miscellaneous fixings

Composite fixing rate for brackets

1 fixing	nr	LQ	0.18
2 fixings	nr	LQ	0.23
3 fixings	nr	LQ	0.40

41

Transformers, switchgear and distribution boards

	Unit	Labour grade	Labour hours
Transformers			
11kV/415 volt, 50Hz three-phase transformer, air cooled, oil filled, skid mounted, cable boxes, fixed to backgrounds			
500kVA	nr	LQ	30.00
800kVA	nr	LQ	40.00
1000kVA	nr	LQ	60.00
1250kVA	nr	LQ	80.00
1500kVA	nr	LQ	100.00
2000kVA	nr	LQ	120.00

	Unit	Labour grade	Labour hours
11kV/415 volt, 50Hz three-phase transformer, air cooled, oil filled, hermetically sealed, skid mounted, cable boxes, fixed to backgrounds			
500kVA	nr	LQ	44.00
800kVA	nr	LQ	56.00
1000kVA	nr	LQ	80.00
1250kVA	nr	LQ	102.00
1500kVA	nr	LQ	124.00
2000kVA	nr	LQ	150.00
11kV/415 volt, 50Hz three-phase transformer, hermetically sealed, silicone impregnated, skid mounted, cable boxes, fixed to backgrounds			
500kVA	nr	LQ	35.00
800kVA	nr	LQ	45.00
1000kVA	nr	LQ	65.00
1250kVA	nr	LQ	85.00
1500kVA	nr	LQ	105.00
2000kVA	nr	LQ	125.00

	Unit	Labour grade	Labour hours
11kV/415 volt, 50Hz three-phase transformer, ventilated case, skid mounted, cable boxes, fixed to backgrounds			
500kVA	nr	LQ	30.00
800kVA	nr	LQ	40.00
1000kVA	nr	LQ	60.00
1250kVA	nr	LQ	80.00
1500kVA	nr	LQ	100.00
2000kVA	nr	LQ	120.00
11kV/415 volt, 50Hz three-phase transformer, cast resin core and coils, skid mounted, cable boxes, fixed to backgrounds			
500kVA	nr	LQ	35.00
800kVA	nr	LQ	45.00
1000kVA	nr	LQ	65.00
1250kVA	nr	LQ	85.00
1500kVA	nr	LQ	105.00
2000kVA	nr	LQ	125.00
11kV/415 volt, 50Hz three-phase transformer, cast resin, ventilated case, skid mounted, cable boxes, fixed to backgrounds			
500kVA	nr	LQ	35.00
800kVA	nr	LQ	45.00
1000kVA	nr	LQ	65.00
1250kVA	nr	LQ	85.00
1500kVA	nr	LQ	105.00
2000kVA	nr	LQ	125.00

	Unit	Labour grade	Labour hours

Busbar trunking

Horizontal mounted busbar trunking, copper conductor bars extruded aluminium housing, earth continuity, fixed to purpose-made supports   2

	Unit	Labour grade	Labour hours
160 Amp	m	LQ	2.88
End cap	nr	LQ	0.20
Angle 90 degrees, flat edge	nr	LQ	1.34
Tee, flat edge	nr	LQ	1.50
End feed unit	nr	LQ	2.50
Centre feed unit	nr	LQ	2.50
Fire barrier	nr	LQ	0.34
Support	nr	LQ	1.00
250 Amp	m	LQ	3.00
End cap	nr	LQ	0.40
Angle 90 degrees, flat edge	nr	LQ	1.34
Tee, flat edge	nr	LQ	1.50
End feed unit	nr	LQ	2.66
Centre feed unit	nr	LQ	2.66
Fire barrier	nr	LQ	0.34
Support	nr	LQ	1.00
315 Amp	m	LQ	3.24
End cap	nr	LQ	0.40
Angle 90 degrees, flat edge	nr	LQ	1.34
Tee, flat edge	nr	LQ	1.50
End feed unit	nr	LQ	2.75
Centre feed unit	nr	LQ	2.75
Fire barrier	nr	LQ	0.50
Support	nr	LQ	1.34

	Unit	Labour grade	Labour hours
400 Amp	m	LQ	3.60
End cap	nr	LQ	0.40
Angle 90 degrees, flat edge	nr	LQ	1.50
Tee, flat edge	nr	LQ	2.00
End feed unit	nr	LQ	3.00
Centre feed unit	nr	LQ	3.00
Fire barrier	nr	LQ	0.50
Support	nr	LQ	1.34
630 Amp	m	LQ	5.04
End cap	nr	LQ	0.80
Angle 90 degrees, flat edge	nr	LQ	2.00
Tee, flat edge	nr	LQ	2.60
End feed unit	nr	LQ	3.66
Centre feed unit	nr	LQ	3.00
Fire barrier	nr	LQ	0.66
Support	nr	LQ	1.50

LV switchboards

Low voltage switchboard Form 4, rear
access fully extendable to take fuse
switches or MCCBs, erected on
prepared foundations, floorstanding,
placed in position

400 Amp rated, busbar chamber
measured as one unit

	Unit	Labour grade	Labour hours
Up to 4 units	nr	LQ	50.00
5 to 7 units	nr	LQ	60.00
8 to 10 units	nr	LQ	70.00
11 to 13 units	nr	LQ	80.00
14 to 15 units	nr	LQ	100.00
16 to 20 units	nr	LQ	120.00

	Unit	Labour grade	Labour hours
800 Amp rated, busbar chamber measured as one unit			
Up to 4 units	nr	LQ	70.00
5 to 7 units	nr	LQ	80.00
8 to 10 units	nr	LQ	90.00
11 to 13 units	nr	LQ	100.00
14 to 15 units	nr	LQ	120.00
16 to 20 units	nr	LQ	140.00
1600 Amp rated, busbar chamber measured as one unit			
Up to 4 units	nr	LQ	80.00
5 to 7 units	nr	LQ	90.00
8 to 10 units	nr	LQ	100.00
11 to 13 units	nr	LQ	110.00
14 to 15 units	nr	LQ	130.00
16 to 20 units	nr	LQ	150.00

LV sub-switchboards

Low voltage sub-distribution boards
Form 4, to take plug-in type fuse
switches, wallmounted, front access,
fixed to backgrounds

	Unit	Labour grade	Labour hours
1 x 400A incomer 12 x 100A outgoing fuse switches complete with interconnections and fuses	nr	LQ	100.00
1 x 800A incomer 12 x 100A outgoing fuse switches complete with interconnections and fuses	nr	LQ	120.00

	Unit	Labour grade	Labour hours
LV switchgear and distribution boards			
4-Way TP and N including 3 M6 TP MCB	nr	LQ	2.97
12-Way TP and N including 8 M6 TP MCB	nr	LQ	3.65
18-Way TP and N including 12 M6 TP MCB	nr	LQ	4.32
Low voltage fuse distribution board, fully insulated and shrouded of welded construction, rust-proofed and painted including rated fuselinks, fixing to backgrounds			
20A, 4-way SP and N	nr	LQ	1.35
20A, 8-way SP and N	nr	LQ	1.89
20A, 12-way SP and N	nr	LQ	2.30
20A, 18-way SP and N	nr	LQ	2.70
32A, 4-way SP and N	nr	LQ	1.62
32A, 8-way SP and N	nr	LQ	2.03
32A, 12-way SP and N	nr	LQ	2.43
32A, 18-way SP and N	nr	LQ	2.88
63A, 4-way SP and N	nr	LQ	2.30
63A, 8-way SP and N	nr	LQ	2.70
63A, 12-way SP and N	nr	LQ	3.38
100A, 2-way SP and N	nr	LQ	1.90
100A, 4-way SP and N	nr	LQ	3.11
100A, 6-way SP and N	nr	LQ	3.38
100A, 8-way SP and N	nr	LQ	4.05
100A, 12-way SP and N	nr	LQ	4.95

	Unit	Labour grade	Labour hours
Wallmounted switch fuse units including fuselinks, fixed to backgrounds			
20A, SP and N	nr	LQ	1.25
32A, SP and N	nr	LQ	1.25
63A, SP and N	nr	LQ	1.50
100A, SP and N	nr	LQ	1.75
150A, SP and N	nr	LQ	2.00
150A, TP and N	nr	LQ	2.50
200A, TP and N	nr	LQ	3.00
Wallmounted isolator switch, padlocked in the OFF or ON position, fixed to backgrounds			
20A, SP and N	nr	LQ	1.15
32A, SP and N	nr	LQ	1.15
63A, SP and N	nr	LQ	1.40
100A, SP and N	nr	LQ	1.65
200A, TP and N	nr	LQ	1.40
100A, TP and N	nr	LQ	1.90
200A, TP and N	nr	LQ	2.90
400A, TP and N	nr	LQ	5.40
630A, TP and N	nr	LQ	5.90
800A, TP and N	nr	LQ	6.90

	Unit	Labour grade	Labour hours

Contactors and starters

Direct on-line starters, masonry
mounted, inside ABS enclosure
including isolator integral to enclosure
to IP65, thermal overload and AC coil,
fixed to backgrounds

	Unit	Labour grade	Labour hours
Up to 1kW	nr	LQ	1.75
1 to 5kW	nr	LQ	2.50

Stella delta starters, masonry mounted
inside ABS enclosure including isolator
integral to enclosure to IP65, thermal
overload and AC coil, fixed to
backgrounds

	Unit	Labour grade	Labour hours
7.5 to 10kW	nr	LQ	4.00
15 to 20kW	nr	LQ	6.00
20 to 25kW	nr	LQ	7.75

415/240 volt contactor relays, plastic
enclosures, fixed to backgrounds

	Unit	Labour grade	Labour hours
20A, SP and N	nr	LQ	1.35
32A, SP and N	nr	LQ	1.35
60A, SP and N	nr	LQ	1.60
20A, TP and N	nr	LQ	1.60
32A, TP and N	nr	LQ	1.60
32A, rectifier, TP and N	nr	LQ	1.70
63A, TP and N	nr	LQ	1.85
63A, rectifier, TP and N	nr	LQ	1.90

42

Luminaires and lamps

	Unit	Labour grade	Labour hours
Fluorescent batten-type luminaire, single tube, fixed to background			
1200mm long, 36 watt	nr	LQ	1.05
1500mm long, 58 watt	nr	LQ	1.05
1800mm long, 70 watt	nr	LQ	1.05
2400mm long, 100 watt	nr	LQ	1.55
Fluorescent batten-type luminaire, twin tube, fixed to background			
1200mm long 2 x 36 watt	nr	LQ	1.35
1500mm long, 2 x 58 watt	nr	LQ	1.35
1800mm long, 2 x 70 watt	nr	LQ	1.35
2400mm long, 2 x 100 watt	nr	LQ	1.85

	Unit	Labour grade	Labour hours
Fluorescent batten-type luminaire, single tube, prismatic diffuser, fixed to background			
1200mm long, 36 watt	nr	LQ	1.21
1500mm long, 58 watt	nr	LQ	1.25
1800mm long, 70 watt	nr	LQ	1.23
2400mm long, 100 watt	nr	LQ	1.75
Fluorescent batten-type luminaire, twin tube, prismatic diffuser, fixed to background			
1200mm long 2 x 36 watt	nr	LQ	1.51
1500mm long, 2 x 58 watt	nr	LQ	1.55
1800mm long, 2 x 70 watt	nr	LQ	1.55
2400mm long, 2 x 100 watt	nr	LQ	2.05
Fluorescent batten-type luminaire, single tube, reflector, fixed to background			
1200mm long, 36 watt	nr	LQ	1.15
1500mm long, 58 watt	nr	LQ	1.20
1800mm long, 70 watt	nr	LQ	1.20
2400mm long, 100 watt	nr	LQ	1.70

	Unit	Labour grade	Labour hours
Fluorescent batten-type luminaire, twin tube, reflector, fixed to background			
1200mm long 2 x 36 watt	nr	LQ	1.45
1500mm long, 2 x 58 watt	nr	LQ	1.50
1800mm long, 2 x 70 watt	nr	LQ	1.50
2400mm long, 2 x 100 watt	nr	LQ	2.00
Fluorescent batten-type luminaire, single tube, conduit suspension, fixed to background			
1200mm long, 36 watt	nr	LQ	1.95
1500mm long, 58 watt	nr	LQ	1.95
1800mm long, 70 watt	nr	LQ	1.95
2400mm long, 100 watt	nr	LQ	2.45
Fluorescent batten-type luminaire, twin tube, conduit suspension, fixed to background			
1200mm long 2 x 36 watt	nr	LQ	1.95
1500mm long, 2 x 58 watt	nr	LQ	1.95
1800mm long, 2 x 70 watt	nr	LQ	1.95
2400mm long, 2 x 100 watt	nr	LQ	2.45
Fluorescent batten-type luminaire, single tube, chain suspension, fixed to background			
1200mm long, 36 watt	nr	LQ	1.71
1500mm long, 58 watt	nr	LQ	1.75
1800mm long, 70 watt	nr	LQ	1.75
2400mm long, 100 watt	nr	LQ	2.25

	Unit	Labour grade	Labour hours
Fluorescent batten-type luminaire, twin tube, chain suspension, fixed to background			
1200mm long 2 x 36 watt	nr	LQ	2.01
1500mm long, 2 x 58 watt	nr	LQ	2.05
1800mm long, 2 x 70 watt	nr	LQ	2.05
2400mm long, 2 x 100 watt	nr	LQ	2.55
Fluorescent batten-type luminaire, single tube, trunking suspension, fixed to background			
1200mm long, 36 watt	nr	LQ	1.01
1500mm long, 58 watt	nr	LQ	1.05
1800mm long, 70 watt	nr	LQ	1.05
2400mm long, 100 watt	nr	LQ	1.45
Fluorescent batten-type luminaire, twin tube, trunking suspension, fixed to background			
1200mm long 2 x 36 watt	nr	LQ	1.26
1500mm long, 2 x 58 watt	nr	LQ	1.30
1800mm long, 2 x 70 watt	nr	LQ	1.30
2400mm long, 2 x 100 watt	nr	LQ	1.70

	Unit	Labour grade	Labour hours

Fluorescent recessed modular
luminaire, twin tube, prismatic diffuser
switch start, side suspension arm, fixed
in false ceiling

	Unit	Labour grade	Labour hours
300 x 1200mm, 2 x 36 watt	nr	LQ	4.25
600 x 600mm, 2 x 40 watt	nr	LQ	4.35
600 x 1200mm, 2 x 36 watt	nr	LQ	4.60
300 x 1200mm, 3 x 36 watt	nr	LQ	4.75
600 x 600mm, 3 x 40 watt	nr	LQ	4.90
600 x 1200mm, 3 x 36 watt	nr	LQ	5.40
300 x 1200mm, 4 x 36 watt	nr	LQ	4.90
600 x 600mm, 4 x 40 watt	nr	LQ	5.20
600 x 1200mm, 4 x 36 watt	nr	LQ	5.70

Waterproof luminaires

Fluorescent luminaire, single tube, dust-
tight jet-proof switch start, glass
reinforced polyester casing, diffuser,
sealing gasket, fixed to background

	Unit	Labour grade	Labour hours
600mm long, 18 watt	nr	LQ	1.04
1200mm long, 36 watt	nr	LQ	1.31
1800mm long, 70 watt	nr	LQ	1.35
2400mm long, 100 watt	nr	LQ	1.35

	Unit	Labour grade	Labour hours
Fluorescent luminaire, twin tube, dust-tight jet-proof switch start, glass reinforced polyester casing, diffuser, sealing gasket, fixed to background			
600mm long, 2 x 18 watt	nr	LQ	1.36
1200mm long, 2 x 36 watt	nr	LQ	1.64
1800mm long, 2 x 70 watt	nr	LQ	1.68
2400mm long, 2 x 100 watt	nr	LQ	1.68

Emergency luminaires

Emergency lighting luminaire, 3-hour duration batteries, surface- mounted polycarbonate body with opal diffuser, fixed to backgrounds

	Unit	Labour grade	Labour hours
368 x 102mm, 8 watt, non-maintained	nr	LQ	0.95

Emergency lighting luminaire bulkhead, 3 hour duration batteries, stoved enamel box, polycarbonate diffuser, fixed to backgrounds

	Unit	Labour grade	Labour hours
368 x 102mm, 8 watt, non-maintained	nr	LQ	0.95
368 x 102mm, 8 watt, maintained	nr	LQ	0.95

Emergency lighting luminaire EXIT sign, 3-hour duration batteries, stoved enamel box, PVC fascia panel, fixed to backgrounds

	Unit	Labour grade	Labour hours
Single-sided EXIT sign, 8 watt	nr	LQ	0.95
Double-sided EXIT sign, 8 watt	nr	LQ	0.95

	Unit	Labour grade	Labour hours

Industrial lighting

High pressure sodium low bay luminaire, steel, powder-coated body, aluminium reflector, integral control gear, wireguard, fixed to background

150 watt	nr	LQ	1.26
250 watt	nr	LQ	1.26
400 watt	nr	LQ	1.26

High pressure sodium high bay luminaire, epoxy covered wiring box, aluminium reflector, integral control gear, wireguard, fixed to background, suspended with chains and hooks

250 watt medium/wide distribution	nr	LQ	1.26
400 watt medium/wide distribution	nr	LQ	1.26

High pressure mercury high bay luminaire, epoxy covered wiring box, aluminium reflector, integral control gear, wireguard, fixed to background, suspended with chains and hooks

250 watt medium/wide distribution	nr	LQ	1.86
400 watt medium/wide distribution	nr	LQ	1.86

	Unit	Labour grade	Labour hours

Floodlighting

High pressure sodium floodlight,
aluminium alloy powder-coated casing,
stainless steel bolts/screws, toughened
safety glass, gear box, fixed to
background

	Unit	Labour grade	Labour hours
250 watt medium/wide distribution	nr	LQ	1.98
400 watt medium/wide distribution	nr	LQ	1.98

External luminaires

High pressure sodium lantern
aluminium canopy for A class roads,
support clear bowl lighting column,
aluminium bracket, cut-out and the like,
fixed to base

	Unit	Labour grade	Labour hours
250 watt, 10 metres high	nr	LQ	5.41
400 watt, 10 metres high	nr	LQ	5.41
250 watt, 12 metres high	nr	LQ	5.41
400 watt, 12 metres high	nr	LQ	5.41

High pressure sodium lantern
aluminium canopy for B class roads,
support clear bowl lighting column,
aluminium bracket, cut-out and the like
fixed to base

	Unit	Labour grade	Labour hours
35 watt, 5 metres high	nr	LQ	4.66

43

Electrical accessories

	Unit	Labour grade	Labour hours
Lighting			
20 Amp one-way 250 volt grade flush-mounted grid switch accessories including phase barriers where required, face plate and metal box to backgrounds requiring fixing			
One-gang switch	nr	LQ	0.40
Two-gang switch	nr	LQ	0.50
Three-gang switch	nr	LQ	0.68
Four-gang switch	nr	LQ	0.78
Six-gang switch	nr	LQ	1.18
Nine-gang switch	nr	LQ	1.93
Twelve-gang switch	nr	LQ	2.18

	Unit	Labour grade	Labour hours

20 Amp one-way 250 volt grade surface-mounted grid switch accessories including face plate and aluminium box, flexible PVC insulated earth continuity conductor between box and face plate, to backgrounds requiring fixing

	Unit	Labour grade	Labour hours
One-gang switch	nr	LQ	0.37
Two-gang switch	nr	LQ	0.47
Three-gang switch	nr	LQ	0.63
Four-gang switch	nr	LQ	0.73
Six-gang switch	nr	LQ	1.15
Nine-gang switch	nr	LQ	1.62
Twelve-gang switch	nr	LQ	2.11

Plug and socket ceiling rose outlet including small pattern conduit box, fixings, bushes, flexible cables EP rubber insulated sheathed cables from plug top to luminaire, 1500mm length of cable, gland and lock nut

	Unit	Labour grade	Labour hours
One gang with plug, 2 pole and earth, 3 core 1.5mm2	nr	LQ	1.25
One gang with plug, 3 pole and earth, 4 core 1.5mm2	nr	LQ	1.35

	Unit	Labour grade	Labour hours

General purpose power

250 volt grade flush-mounted
accessories including back boxes, fixed
to backgrounds

	Unit	Labour grade	Labour hours
13 Amp, 1 gang switched socket outlet	nr	LQ	0.55
13 Amp, DP, switch fused connection	nr	LQ	0.50

250 volt grade surface mounted
accessories, including back boxes,
flexible PVC insulated earth continuity
conductor between box and face plate,
to backgrounds requiring fixing

	Unit	Labour grade	Labour hours
13 Amp, 1 gang switched socket outlet	nr	LQ	0.55

250 volt grade flush-mounted
accessories, skirting trunking including
back boxes, fixed to backgrounds

	Unit	Labour grade	Labour hours
13 Amp, 2 gang switched socket outlet	nr	LQ	0.50

250 volt grade flush-mounted
accessories, bench trunking including
back boxes, fixed to backgrounds

	Unit	Labour grade	Labour hours
13 Amp, 2 gang switched socket outlet	nr	LQ	0.50

	Unit	Labour grade	Labour hours

Telecommunications

White plastic flush-mounted telephone
outlet including back box, fixed to
backgrounds

| | nr | LQ | 0.40 |

Radio

White plastic flush-mounted FM/AM
outlet including back box, fixed to
backgrounds

| | nr | LQ | 0.40 |

Television

White plastic flush mounted television
outlet including back box, fixed to
backgrounds

| | nr | LQ | 0.40 |

44

Fire detection and alarms

	Unit	Labour grade	Labour hours
Fire alarm fittings, surface mounted including standard besa box, connections, drilling, fixed to backgrounds			
Open circuit break-glass units	nr	LQ	0.55
Fire alarm sounder	nr	LQ	0.65
Optical smoke detector including addressable base	nr	LQ	1.00
Fixed temperature heat detector including addressable base	nr	LQ	1.00
Loop isolators	nr	LQ	0.50

45

Earthing and bonding

	Unit	Labour grade	Labour hours
Taping at 1000mm centres including dressing			
25 x 3mm horizontal bare aluminium tape, clipped/screwed, to backgrounds	m	LQ	0.70
25 x 3mm horizontal bare copper tape, clipped/screwed, to backgrounds	m	LQ	0.72
25 x 3mm horizontal bare PVC-U aluminium tape, clipped/screwed, to backgrounds	m	LQ	0.72
25 x 3mm horizontal bare PVC-U copper tape, clipped/screwed, to backgrounds	m	LQ	0.74
25 x 3mm vertical bare aluminium tape, clipped/screwed, to backgrounds	m	LQ	0.75
25 x 3mm vertical bare copper tape, clipped/screwed, to backgrounds	m	LQ	0.77

	Unit	Labour grade	Labour hours
25 x 3mm vertical PVC-U aluminium tape, clipped/screwed, to backgrounds	m	LQ	0.77
25 x 3mm vertical PVC-U copper tape, clipped/screwed, to backgrounds	m	LQ	0.79

Clamps

	Unit	Labour grade	Labour hours
25 x 3mm aluminium square clamp	nr	LQ	0.50
25 x 3mm gunmetal square clamp	nr	LQ	0.50

Bonds

	Unit	Labour grade	Labour hours
25 x 3mm aluminium bond to metalwork	nr	LQ	0.50
25 x 3mm gunmetal bond to metalwork	nr	LQ	0.50
25 x 3mm aluminium RWP bond to metalwork	nr	LQ	0.50

Test points

	Unit	Labour grade	Labour hours
25 x 3mm bimetallic clamp	nr	LQ	0.50
25 x 3mm gunmetal oblong test clamp	nr	LQ	0.50

Earth rod clamps

	Unit	Labour grade	Labour hours
25 x 3mm gunmetal clamp	nr	LQ	0.50

	Unit	Labour grade	Labour hours
16mm/16-70mm gunmetal GUV clamp	nr	LQ	0.50
8mm/16-70mm gunmetal G clamp	nr	LQ	0.21

Earthing rods

	Unit	Labour grade	Labour hours
1200 x 16mm copperband rod including coupling, driven	nr	LQ	0.50
2400 x 16mm copperband rod including coupling, driven	nr	LQ	1.00

Earthing mats/plates

	Unit	Labour grade	Labour hours
600 x 600 x 3mm copper lattice mat laid in prepared excavation	nr	LQ	1.00
900 x 900 x 3mm copper lattice mat laid in prepared excavation	nr	LQ	1.25
900 x 900 x 3mm solid copper mat laid in prepared excavation	nr	LQ	1.25

Earth tape/cable

	Unit	Labour grade	Labour hours
25 x 3mm bare copper tape laid in trench	m	LQ	0.15
25 x 3mm PVC-U copper tape laid in trench	m	LQ	0.15

	Unit	Labour grade	Labour hours

Earth bars, fixed to backgrounds

	Unit	Labour grade	Labour hours
50 x 6 x 600mm hard drawn tinned copper earth bar	nr	LQ	1.00
50 x 6 x 1000mm single disconnect link	nr	LQ	1.65
50 x 6 x 1000mm twin disconnect link	nr	LQ	2.25
50 x 6 x 1500mm twin disconnect link	nr	LQ	2.50

46

Cables and wiring

	Unit	Labour grade	Labour hours

Multicore armoured PVC insulated cables (copper/PVC/SWA/PVC)

Clipped to surfaces, 2 cores

	Unit	Labour grade	Labour hours
1.5mm2	m	LQ	0.32
2.5mm2	m	LQ	0.32
4mm2	m	LQ	0.34
6mm2	m	LQ	0.34
10mm2	m	LQ	0.37
16mm2	m	LQ	0.37
25mm2	m	LQ	0.43
35mm2	m	LQ	0.46
50mm2	m	LQ	0.50
70mm2	m	LQ	0.50
95mm2	m	LQ	0.53
120mm2	m	LQ	0.59

	Unit	Labour grade	Labour hours

Clipped to tray, 2 cores

	Unit	Labour grade	Labour hours
1.5mm2	m	LQ	0.20
2.5mm2	m	LQ	0.20
4mm2	m	LQ	0.22
6mm2	m	LQ	0.22
10mm2	m	LQ	0.25
16mm2	m	LQ	0.25
25mm2	m	LQ	0.28
35mm2	m	LQ	0.31
50mm2	m	LQ	0.35
70mm2	m	LQ	0.35
95mm2	m	LQ	0.38
120mm2	m	LQ	0.44

Laid in trenches or drawn through ducts, 2 cores

	Unit	Labour grade	Labour hours
1.5mm2	m	LQ	0.14
2.5mm2	m	LQ	0.14
4mm2	m	LQ	0.16
6mm2	m	LQ	0.16
10mm2	m	LQ	0.18
16mm2	m	LQ	0.18
25mm2	m	LQ	0.21
35mm2	m	LQ	0.23
50mm2	m	LQ	0.27
70mm2	m	LQ	0.27
95mm2	m	LQ	0.30
120mm2	m	LQ	0.33

	Unit	Labour grade	Labour hours
Clipped to surfaces, 3 cores			
1.5mm2	m	LQ	0.32
2.5mm2	m	LQ	0.32
4mm2	m	LQ	0.34
6mm2	m	LQ	0.34
10mm2	m	LQ	0.38
16mm2	m	LQ	0.40
25mm2	m	LQ	0.48
35mm2	m	LQ	0.50
50mm2	m	LQ	0.53
70mm2	m	LQ	0.57
95mm2	m	LQ	0.57
120mm2	m	LQ	0.69
Clipped to tray, 3 cores			
1.5mm2	m	LQ	0.20
2.5mm2	m	LQ	0.20
4mm2	m	LQ	0.22
6mm2	m	LQ	0.22
10mm2	m	LQ	0.26
16mm2	m	LQ	0.28
25mm2	m	LQ	0.33
35mm2	m	LQ	0.35
50mm2	m	LQ	0.38
70mm2	m	LQ	0.44
95mm2	m	LQ	0.44
120mm2	m	LQ	0.54

	Unit	Labour grade	Labour hours
Laid in trenches or drawn through ducts, 3 cores			
1.5mm2	m	LQ	0.14
2.5mm2	m	LQ	0.14
4mm2	m	LQ	0.16
6mm2	m	LQ	0.16
10mm2	m	LQ	0.20
16mm2	m	LQ	0.21
25mm2	m	LQ	0.25
35mm2	m	LQ	0.27
50mm2	m	LQ	0.30
70mm2	m	LQ	0.33
95mm2	m	LQ	0.33
120mm2	m	LQ	0.43
Clipped to surfaces, 4 cores			
1.5mm2	m	LQ	0.32
2.5mm2	m	LQ	0.34
4mm2	m	LQ	0.34
6mm2	m	LQ	0.37
10mm2	m	LQ	0.43
16mm2	m	LQ	0.46
25mm2	m	LQ	0.50
35mm2	m	LQ	0.50
50mm2	m	LQ	0.53
70mm2	m	LQ	0.57
95mm2	m	LQ	0.69
120mm2	m	LQ	0.77

	Unit	Labour grade	Labour hours
Laid in trenches or drawn through ducts, 4 cores			
1.5mm2	m	LQ	0.14
2.5mm2	m	LQ	0.16
4mm2	m	LQ	0.16
6mm2	m	LQ	0.21
10mm2	m	LQ	0.20
16mm2	m	LQ	0.23
25mm2	m	LQ	0.27
35mm2	m	LQ	0.27
50mm2	m	LQ	0.30
70mm2	m	LQ	0.33
95mm2	m	LQ	0.43
120mm2	m	LQ	0.50
Terminations for PVC insulated armoured cable including connections, 2 cores			
1.5mm2	nr	LQ	0.66
2.5mm2	nr	LQ	0.66
4mm2	nr	LQ	0.66
6mm2	nr	LQ	0.88
10mm2	nr	LQ	0.99
16mm2	nr	LQ	1.19
25mm2	nr	LQ	1.39
35mm2	nr	LQ	1.59
50mm2	nr	LQ	2.00
70mm2	nr	LQ	2.20
95mm2	nr	LQ	2.63
120mm2	nr	LQ	2.88

	Unit	Labour grade	Labour hours

Terminations for PVC insulated
armoured cable including connections,
3 cores

1.5mm2	nr	LQ	0.75
2.5mm2	nr	LQ	0.75
4mm2	nr	LQ	0.75
6mm2	nr	LQ	0.92
10mm2	nr	LQ	1.09
16mm2	nr	LQ	1.39
25mm2	nr	LQ	1.69
35mm2	nr	LQ	2.20
50mm2	nr	LQ	2.50
70mm2	nr	LQ	3.06
95mm2	nr	LQ	3.36
120mm2	nr	LQ	3.97

Terminations for PVC insulated
armoured cable including connections,
4 cores

1.5mm2	nr	LQ	0.83
2.5mm2	nr	LQ	0.83
4mm2	nr	LQ	1.00
6mm2	nr	LQ	1.00
10mm2	nr	LQ	1.19
16mm2	nr	LQ	1.59
25mm2	nr	LQ	2.20
35mm2	nr	LQ	2.60
50mm2	nr	LQ	3.26
70mm2	nr	LQ	3.66
95mm2	nr	LQ	4.07
120mm2	nr	LQ	4.97

	Unit	Labour grade	Labour hours

**Mineral insulated cables (MICC)
fixed to backgrounds, including
bends and dressing**

Light duty cables, bare copper sheath,
2 cores

1mm2	m	LQ	0.17
1.5mm2	m	LQ	0.18
2.5mm2	m	LQ	0.20
4mm2	m	LQ	0.22

Light duty cables, bare copper sheath,
3 cores

1mm2	m	LQ	0.18
1.5mm2	m	LQ	0.20
2.5mm2	m	LQ	0.22

Light duty cables, bare copper sheath,
4 cores

1mm2	m	LQ	0.20
1.5mm2	m	LQ	0.21
2.5mm2	m	LQ	0.24

Light duty cables, bare copper sheath,
7 cores

1mm2	m	LQ	0.24
1.5mm2	m	LQ	0.25
2.5mm2	m	LQ	0.28

	Unit	Labour grade	Labour hours
Light duty cables, PVC sheath, 2 cores			
1mm2	m	LQ	0.18
1.5mm2	m	LQ	0.19
2.5mm2	m	LQ	0.21
4mm2	m	LQ	0.22
Light duty cables, PVC sheath, 3 cores			
1mm2	m	LQ	0.90
1.5mm2	m	LQ	0.21
2.5mm2	m	LQ	0.23
Light duty cables, bare copper sheath, 4 cores			
1mm2	m	LQ	0.21
1.5mm2	m	LQ	0.22
2.5mm2	m	LQ	0.25
Light duty cables, bare copper sheath, 7 cores			
1mm2	m	LQ	0.26
1.5mm2	m	LQ	0.27
2.5mm2	m	LQ	0.30

	Unit	Labour grade	Labour hours

Light duty cables, bare copper, 2 cores

1.5mm2	m	LQ	0.19
2.5mm2	m	LQ	0.21
4mm2	m	LQ	0.24

Light duty cables, bare copper, 3 cores

1.5mm2	m	LQ	0.20
2.5mm2	m	LQ	0.22
4mm2	m	LQ	0.26

Light duty cables, bare copper, 3 cores

1.5mm2	m	LQ	0.20
2.5mm2	m	LQ	0.22
4mm2	m	LQ	0.26

Light duty cables, bare copper, 4 cores

1.5mm2	m	LQ	0.21
2.5mm2	m	LQ	0.23
4mm2	m	LQ	0.26

Light duty cables, bare copper, 7 cores

1.5mm2	m	LQ	0.24
2.5mm2	m	LQ	0.28

	Unit	Labour grade	Labour hours
Light duty cables, bare copper, 12 cores			
2.5mm2	m	LQ	0.35
Light duty cables, bare copper, 19 cores			
1.5mm2	m	LQ	0.37
Heavy duty cables, bare copper, 1 core			
6mm2	m	LQ	0.19
10mm2	m	LQ	0.21
16mm2	m	LQ	0.22
25mm2	m	LQ	0.26
35mm2	m	LQ	0.30
50mm2	m	LQ	0.33
70mm2	m	LQ	0.35
95mm2	m	LQ	0.39
120mm2	m	LQ	0.44
150mm2	m	LQ	0.46
Heavy duty cables, bare copper, 2 cores			
1.5mm2	m	LQ	0.19
2.5mm2	m	LQ	0.21
4mm2	m	LQ	0.24
6mm2	m	LQ	0.27
10mm2	m	LQ	0.29
16mm2	m	LQ	0.33
25mm2	m	LQ	0.37

	Unit	Labour grade	Labour hours
Heavy duty cables, bare copper, 3 cores			
1.5mm2	m	LQ	0.20
2.5mm2	m	LQ	0.22
4mm2	m	LQ	0.26
6mm2	m	LQ	0.29
10mm2	m	LQ	0.32
16mm2	m	LQ	0.35
25mm2	m	LQ	0.41
Heavy duty cables, bare copper, 4 cores			
1.5mm2	m	LQ	0.21
2.5mm2	m	LQ	0.23
4mm2	m	LQ	0.29
6mm2	m	LQ	0.31
10mm2	m	LQ	0.35
16mm2	m	LQ	0.39
25mm2	m	LQ	0.46
Heavy duty cables, bare copper, 7 cores			
1.5mm2	m	LQ	0.24
2.5mm2	m	LQ	0.28
Heavy duty cables, bare copper, 12 cores			
2.5mm2	m	LQ	0.28

	Unit	Labour grade	Labour hours
Heavy duty cables, bare copper, 19 cores			
2.5mm2	m	LQ	0.37
Heavy duty cables, PVC sheath, 1 core			
6mm2	m	LQ	0.20
10mm2	m	LQ	0.22
16mm2	m	LQ	0.23
25mm2	m	LQ	0.27
35mm2	m	LQ	0.32
50mm2	m	LQ	0.35
70mm2	m	LQ	0.37
95mm2	m	LQ	0.41
120mm2	m	LQ	0.46
150mm2	m	LQ	0.50
Heavy duty cables, PVC sheath, 2 cores			
1.5mm2	m	LQ	0.20
2.5mm2	m	LQ	0.22
4mm2	m	LQ	0.25
6mm2	m	LQ	0.28
10mm2	m	LQ	0.30
16mm2	m	LQ	0.34
25mm2	m	LQ	0.38

	Unit	Labour grade	Labour hours
Heavy duty cables, PVC sheath, 3 cores			
1.5mm2	m	LQ	0.21
2.5mm2	m	LQ	0.23
4mm2	m	LQ	0.27
6mm2	m	LQ	0.30
10mm2	m	LQ	0.33
16mm2	m	LQ	0.37
25mm2	m	LQ	0.48
Heavy duty cables, PVC sheath, 4 cores			
1.5mm2	m	LQ	0.22
2.5mm2	m	LQ	0.24
4mm2	m	LQ	0.30
6mm2	m	LQ	0.32
10mm2	m	LQ	0.37
16mm2	m	LQ	0.42
25mm2	m	LQ	0.49
Heavy duty cables, PVC sheath, 7 cores			
1.5mm2	m	LQ	0.25
2.5mm2	m	LQ	0.28
Heavy duty cables, PVC sheath, 12 cores			
2.5mm2	m	LQ	0.38
Heavy duty cables, PVC sheath, 19 cores			
1.5mm2	m	LQ	0.40

	Unit	Labour grade	Labour hours
Terminations for MICC insulated cables including connections, 1 core			
10mm2	m	LQ	0.37
16mm2	m	LQ	0.46
25mm2	m	LQ	0.57
35mm2	m	LQ	0.66
50mm2	m	LQ	0.89
70mm2	m	LQ	1.00
95mm2	m	LQ	1.10
120mm2	m	LQ	1.35
150mm2	m	LQ	1.55
Terminations for MICC insulated cables including connections, 2 cores			
1.0mm2	m	LQ	0.31
1.5mm2	m	LQ	0.33
2.5mm2	m	LQ	0.36
4mm2	m	LQ	0.38
6mm2	m	LQ	0.40
10mm2	m	LQ	0.44
16mm2	m	LQ	0.53
25mm2	m	LQ	0.65

	Unit	Labour grade	Labour hours
Terminations for MICC insulated cables including connections, 3 cores			
1.0mm2	m	LQ	0.36
1.5mm2	m	LQ	0.38
2.5mm2	m	LQ	0.40
4mm2	m	LQ	0.42
6mm2	m	LQ	0.44
10mm2	m	LQ	0.50
16mm2	m	LQ	0.62
25mm2	m	LQ	0.79
Terminations for MICC insulated cables including connections, 4 cores			
1.0mm2	m	LQ	0.39
1.5mm2	m	LQ	0.41
2.5mm2	m	LQ	0.44
4mm2	m	LQ	0.46
6mm2	m	LQ	0.50
10mm2	m	LQ	0.58
16mm2	m	LQ	0.75
25mm2	m	LQ	0.93
Terminations for MICC insulated cables including connections, 7 cores			
1.0mm2	m	LQ	0.48
1.5mm2	m	LQ	0.51
2.5mm2	m	LQ	0.57

	Unit	Labour grade	Labour hours
Terminations for MICC insulated cables including connections, 12 cores			
2.5mm2	m	LQ	0.85
Terminations for MICC insulated cables including connections, 19 cores			
2.5mm2	m	LQ	1.00

Single core PVC insulated PVC sheathed cable

Clipped to surfaces

	Unit	Labour grade	Labour hours
1.0mm2	m	LQ	0.03
1.5mm2	m	LQ	0.03
2.5mm2	m	LQ	0.03
4mm2	m	LQ	0.05
6mm2	m	LQ	0.06
10mm2	m	LQ	0.06
16mm2	m	LQ	0.06
25mm2	m	LQ	0.07
35mm2	m	LQ	0.09
50mm2	m	LQ	0.10
70mm2	m	LQ	0.13
95mm2	m	LQ	0.16

	Unit	Labour grade	Labour hours
Fixed in chases, covered with galvanised or PVC sheath			
1.0mm2	nr	LQ	0.30
1.5mm2	nr	LQ	0.30
2.5mm2	nr	LQ	0.30
4mm2	nr	LQ	0.33
6mm2	nr	LQ	0.33
10mm2	nr	LQ	0.40
16mm2	nr	LQ	0.50

Single core PVC insulated cable, non-armoured non-sheathed (6491X singles)

Drawn in conduit

	Unit	Labour grade	Labour hours
1.0mm2	nr	LQ	0.03
1.5mm2	nr	LQ	0.03
2.5mm2	nr	LQ	0.03
4mm2	nr	LQ	0.04
6mm2	nr	LQ	0.05
10mm2	nr	LQ	0.05
16mm2	nr	LQ	0.05

	Unit	Labour grade	Labour hours
Installed in trunking			
1.0mm2	nr	LQ	0.02
1.5mm2	nr	LQ	0.02
2.5mm2	nr	LQ	0.02
4mm2	nr	LQ	0.04
6mm2	nr	LQ	0.05
10mm2	nr	LQ	0.05
16mm2	nr	LQ	0.05
25mm2	m	LQ	0.06
35mm2	m	LQ	0.08
50mm2	m	LQ	0.10
70mm2	m	LQ	0.12
95mm2	m	LQ	0.15
120mm2	m	LQ	0.18
150mm2	m	LQ	0.21

PART SIX

GENERAL INFORMATION

47

Measurement data

The metric system

Linear

1 centimetre (cm)	= 10 millimetres (mm)
1 decimetre (dm)	= 10 centimetres (cm)
1 metre (m)	= 10 decimetres (dm)
1 kilometre (km)	= 1000 metres (m)

Area

100 sq millimetres	= 1 sq centimetre
100 sq centimetres	= 1 sq decimetre
100 sq decimetres	= 1 sq metre
1000 sq metres	= 1 hectare

Capacity

1 millilitre (ml)	= 1 cubic centimetre (cm3)
1 centilitre (cl)	= 10 millilitres (ml)
1 decilitre (dl)	= 10 centilitres (cl)
1 litre (l)	= 10 decilitres (dl)

Weight

1 centigram (cg)	= 10 milligrams (mg)
1 decigram (dg)	= 10 centigrams (cg)
1 gram (g)	= 10 decigrams (dg)
1 decagram (dag)	= 10 grams (g)
1 hectogram (hg)	= 10 decagrams (dag)
1 kilogram (kg)	= 10 hectogram (hg)

Conversion equivalents (imperial/metric)

Length

1 inch	=	25.4 mm
1 foot	=	304.8 mm
1 yard	=	914.4 mm
1 yard	=	0.9144 m
1 mile	=	1609.34 m

Area

1 sq inch	=	645.16 sq mm
1 sq ft	=	0.092903 sq m
1 sq yard	=	0.8361 sq m
1 acre	=	4840 sq yards
1 acre	=	2.471 hectares

Liquid

1 lb water	=	0.454 litres
1 pint	=	0.568 litres
1 gallon	=	4.546 litres

Horse-power

1 hp	=	746 watts
1 hp	=	0.746 kW
1 hp	=	33,000 ft.lb/min

Weight

1 lb	=	0.4536 kg
1 cwt	=	50.8 kg
1 ton	=	1016.1 kg

Conversion equivalents (metric/imperial)

Length

1 mm	=	0.03937 inches
1 centimetre	=	0.3937 inches
1 metre	=	1.094 yards
1 metre	=	3.282 ft
1 kilometre	=	0.621373 miles

Area

1 sq mm	=	0.00155 sq in
1 sq m	=	10.764 sq ft
1 sq m	=	1.196 sq yards
1 acre	=	4046.86 sq m
1 hectare	=	0.404686 acres

Liquid

1 litre	=	2.202 lbs
1 litre	=	1.76 pints
1 litre	=	0.22 gallons

Horse-power

1 watt	=	0.00134 hp
1 kw	=	134 hp
1 hp	=	0759 kg m/s

Weight

1 kg	=	2.205 lbs
1 kg	=	0.01968 cwt
1 kg	=	0.000984 ton

Temperature equivalents

In order to convert Fahrenheit to Celsius deduct 32 and multiply by 5/9.
To convert Celsius to Fahrenheit multiply by 9/5 and add 32.

Fahrenheit	Celsius
230	110.0
220	104.4
210	98.9
200	93.3
190	87.8
180	82.2
170	76.7
160	71.1
150	65.6
140	60.0
130	54.4
120	48.9
110	43.3
90	32.2
80	26.7
70	21.1
60	15.6
50	10.0
40	4.4
30	-1.1
20	-6.7
10	-12.2
0	-17.8

Areas and volumes

Figure	Area	Perimeter
Rectangle	Length x breadth	Sum of sides
Triangle	Base x half of perpendicular height	Sum of sides
Quadrilateral	Sum of areas of contained triangles	Sum of sides

Trapezoidal	Sum of areas of contained triangles	Sum of sides
Trapezium	Half of sum of parallel sides x perpendicular height	Sum of sides
Parallelogram	Base x perpendicular height	Sum of sides
Regular polygon	Half sum of sides x half internal diameter	Sum of sides
Circle	pi x radius^2	pi x diameter or pi x 2 x radius

Figure	Surface area	Volume
Cylinder	pi x 2 x radius x length (curved surface only)	pi x radius^2 x length div3
Sphere	pi x diameter^2	Diameter^3 x 0.5236
Pyramid	Half base perimeter x sloping height plus area at base	Base area x vertical height divided by 3

Paper sizes

A0	841 x 1189mm
A1	594 x 841mm
A2	420 x 594mm
A3	297 x 420mm
A4	210 x 297mm
A5	148 x 210mm
A6	105 x 148mm
Imperial	30 x 22 inches
Super Royal	27 x 19 inches
Royal	24 x 19 inches
Half Imperial	15 x 22 inches
Foolscap	17 x 13.5 inches

48

Useful addresses

Architects Registration Council of the United Kingdom
73 Hallam Street
London W1N 6EE
(0171-580 5861)

Association of Consulting Engineers
Alliance House
12 Caxton Street
London SW1H 1QL
(0171-222 6557)

Brick Development Association
Woodside House
Winkfield
Windsor
Berkshire SL4 2DP
(01344 885651)

British Board of Agreement
PO Box
195, Bucknall's Lane
Watford
Hertfordshire WD2 7NG
(01923 670844)

British Computer Society
1 Sansord Street
Swindon
Wiltshire SN1 1HJ
(01793 417417)

British Property Federation
35 Catherine Place
London SW1E 6DY
(0171-828 0111)

British Standards Institution
389 Chiswick High Street
London W4 4AL
(0171-629 9000)

British Steel Plc
9 Albert Embankment
London SE1 7SN
(0171-735 7654)

British Woodworking Federation
82 New Cavendish Street
London W1M 8AD
(0171-872 8210)

Builders Merchants Federation
15 Soho Square
London W1V 5FB
(0171-439 1753)

Building Centre
26 Store Street
London WC1E 7BT
(0171-637 1022/8361)
(Information: 0344 884999)

Building Employers Confederation
18 Duchess Mews
London W1
(0171-636 3891)

Building Research Establishment
Bucknalls Lane
Watford
Hertfordshire WD2 7JR
(01923 664000)

Central Office of Information
Hercules Road
London SE1 7DU
(0171-928 2345)

Chartered Institute of Arbitrators
24 Angle Gate
London EC1
(0171-837 4483)

Chartered Institute of Building
Englemere,
Kings Ride
Ascot
Berkshire SL5 8BJ
(01344 630700)

Confederation of British Industry
Centre Point
103 New Oxford Street
London WC1
(0171-379 7400)

Electrical Contractors Association
ESCA House
34 Palace Court
Bayswater
London W2
(0171-229 1266)

Federation of Building Sub-Contractors
82 New Cavendish Street
London W1M 8AD
(0171-580 5588)

Federation of Master Builders
14 Great James Street
London WC1N 2DP
(0171-242 7583)

Glass Manufacturers Federation
19 Portland Place
London W1N 4BH
(0171-580 6952)

Heating and Ventilation Contractors Association
ESCA House
34 Palace Court
London W2 4JG
(0171-229 2488)

Housing Corporation
149 Tottenham Court Road
London W1P 0BN
(0171-393 2000)

Institute of Mechanical Engineers
1 Birdcage Walk
London SW1H 9JJ
(0171-222 7899)

Institute of Plumbing
64 Station Lane
Hornchurch
Essex RN12 6NB
(017108 472791)

Institution of Civil Engineers
1-7 Great George Street
London SW1P 3AA
(0171-222 7722)

Institution of Civil Engineering Surveyors
26 Market Street
Altrincham
Cheshire WA14 1PF
(0161-928 8074)

Institution of Electrical Engineers
2 Savoy Place
London WC2R 0BL
(0171-240 1871)

Institution of Structural Engineers
11 Upper Belgrave Street
London SW1X 8BH
(0171-235 4535/6841)

Iron and Steel Trades Confederation
Swinton House
324 Grays Inn Road
London WC1X 8DD
(0171-837 6691)

Joint Industry Board for the Electrical Contracting Industry
Kingswood House
47/51 Sidcup Hill
Sidcup
Kent DA14 6HJ
(0181 302 0031)

National Association of Local Councils
109 Great Russell Street
London WC1B 3LD
(0171-637 1865)

National Association of Plumbing, Heating and Mechanical Services Contractors
6 Gate Street
London WC2A 3HX
(0171-405 2678)

National Association of Scaffolding Contractors
18 Mansfield Street
London W1M 9FG
(0171-580 558)

National Association of Shopfitters
NAS House
411 Limpsfield Road
The Green
Warlingham
Surrey CR3 9HA
(01883 624961)

National Building Specification
Mansion House Chambers
The Close
Newcastle upon Tyne NE1 30E
(0191-232 9594)

National Computing Centre
Oxford Road
Manchester M1 7ED
(0161-242 2100)

National Council of Building Material Producers
26 Store Street
London WC1E 7BT
(0171-323 3770)

National Federation of Demolition Contractors
1A New Road
The Causeway
Staines
Middlesex TW18 3DH
(0171-404 4020)

National Housing Federation
175 Grays Inn Road
London WC1X 8UP
(0171-278 6571)

National Federation of Painting and Decorating Contractors
18 Mansfield Street
London W1M 9FG
(0171-580 5588)

National Federation of Plastering Contractors
82 New Cavendish Street
London W1M 8AD
(0171-580 5588)

National Federation of Roofing Contractors
24 Weymouth Street
London W1N 4LX
(0171-436 0387)

National Joint Council for Felt Roofing Contracting Industry
Fields House
Gower Road
West Sussex RH16 4PL
(01444 440027)

Royal Institute of British Architects
66 Portland Place
London W1N 4AD
(0171-580 5533)

Royal Institute of Chartered Surveyors
12 Great George Street
London SW1Y 5AG
(0171-222 7000)

The Brick Development Association
Woodside House
Windsor
Berkshire SL4 2DX
(013447 885651)

Town and Country Planning Association
17 Carlton House Terrace
London SW1Y 5AS
(0171-930 8903/5)

Water Authorities Association
1 Queen Anns Gate
London SW1H 9BT
(0171-957 4567)

Welsh Development Agency
Treforest Industrial Estate
Pontypridd
Glamorgan CS37 5UT
(01345 775577)

Welsh Office
Cathays Park
Cardiff CF1 3NQ
(01222 825111)

Index

Air bricks, 58-59
Aluminium sheeting, 74, 87-88
Angle of repose, 4, 272
Anti-sun glass, 207-210
Architraves
  hardwood, 115
  softwood, 114
Artex paint, 227-228
Asphalt work
  cold, 352
  flooring, 69
  hot, 352
  tanking, 68

Bars, 135-137
Base boarding, 158
Beams
  laminated, 107-108
  steel, 125-128, 131
Bituminous paint, 402-403
Blockboard
  casings, 116-119
  linings, 116-119
Blocks
  per m2, 41, 362
Blockwork, 49-53, 372-373
Boilers, 498-500
Bolts, 132-134, 389-390
Bonds, 554
Bricks
  pavings, 236
  per m2, 40-41, 360-361
Brick reinforcement, 57
Brickwork
  common, 45-47, 360-367
  engineering, 47-48, 370-371
  facing, 49, 368-369

Built-up roofing, 93-95
Bulbs, 434

Cables, 557-561, 563-569,
  572-574
Cables and wiring, 557-574
Cable ducts, 342-343
Cable trays, 524-526
Calorifier, 497
Cappings, 37
Carbon steel
  fittings, 473-475
  pipework, 473-475
Carpet, 151
Carpet tiles, 150
Casings
  blockboard, 116-119
  chipboard, 116-119
  hardboard, 116-119
  insulation board, 117-118
  plywood, 117-119
Clamps, 554
Conduit and cable trunking,
  519-527
Conduits
  boxes, 522
  PVC, 520-521
  steel, 519-520
Connectors, 391-392
Containment, 522
Contactors, 537
Copper
  fittings, 451-455
  pipework, 451-455
Cork tiling, 149
Cover plates, 460, 476
Culverts, 344-345
Cylinders, 193, 496-497

Cement-bound aggregate, 350
Ceramic tiling
  floor, 147
  wall, 173-174
Chainlink fencing, 239, 413-414
Channels
  concrete, 238, 356-357
Chestnut fencing, 238-239, 410
Chimney pots, 58
Chipboard
  casings, 116-119
  linings, 116-119
Cisterns, 175
Cladding, 96
Columns, 125-127
Concrete
  finishes, 19, 236, 307-309
  mixes, 21-22, 290
  pavings, 352
  placing, 294-297
  ready mixed, 24-25
  site mixed, 25-26, 235, 293
  sundries, 29-31
Copings, 37
Copper
  fittings, 451-455
  pipework, 182-185, 451-455
  sheeting, 74, 88-90
Cover plates, 460, 476
Cylinders, 496-497

Damp-proof course, 43, 54-56,
  364
Damp-proofing, 406
Decking, 108, 387
Demolition, 267-268
Disposal, 16-17, 278-279
Distribution boards, 535-536
Ditching, 341-342, 440
Door frames and linings, 113-114
Doors, 111-112, 113

Double glazing, 210
Double handling, 280
Dowels, 306-307
Drainage
  generally, 243-256, 309-345
  beds and coverings, 249-250
  fillings, 328-332
  manholes, 256-258, 336-340
  pipework, 251-255, 313-328
  trenches, 245-248

Earth
  bars, 556
  cable, 555
  mats, 555
  plates, 555
  rods, 555
  tape, 553-554
Earthing and bonding, 553-556
Earth rod clamps, 554-555
Earthwork support, 12-16
Eaves, 103, 104-106
Edgings, 237
Emulsion paint, 222-223, 398-400
Excavation, 7-9, 9-12, 276-278
Excavation and filling, 3-19,
  271-285
Expansion devices, 489-490

Fans, 508-511
Fascias, 103, 104-106
Felt roofing, 75
Fencing
  chainlink, 239, 413-414
  chestnut, 238-239, 410
  hurdle, 241
  palisade, 242
  panel, 240
  post and rail, 239, 410-411
  post and wire, 240-241,
    411-412, 415-416

Fibre-covered slating, 78-791
Filling, 17-18, 243, 281-285, 310-311, 436
Fire dectection and alarms, 549
Fittings and fastenings
  bolts, 389-390
  connectors, 391-392
  screws, 388-389
  spikes, 388
  straps, 388
Floodlighting, 544
Floor finishes, 139-152
Flooring
  asphalt, 69
  screeds, 139
  softwood, 109
  terrazzo, 152
  tiling, 139, 146-150
Flush doors, 111-112
Formwork, 23, 26-28, 235, 291, 297-301
French drains, 340
Fuel consumption, 5-6, 273-274

Gabions, 419-420
Gates, 416-417
Gauges, 491
General information, 575-589
Geotextiles, 287-288
Glass
  anti-sun, 207-210
  cast, 205-207
  clear, 197, 199-201
  double glazing, 210
  float, 197, 199-201
  patterned, 197, 201-203
  rough cast, 203-205
Glazing, 197-210
Glazing beads, 115-116

Granolithic, 142-143
Granular material, 234-235, 349
Gravel, 236
Greenheart, 384-385, 387
Grouting, 308
Gutters, 103, 104-106

Hardboard
  casings, 116, 117-119
  linings, 116, 117-119
Hardcore, 234-235, 350
Hardwood, 150, 384-385
Heating and cooling system, 467-505
Hedges, 433
Hurdle-fencing, 241

Immersion heater, 498
Industrial lighting, 543
Insulation, 192
Insulator board
  casings, 117-119
  linings, 117-119
Ironmongery, 120-124

Joints
  contraction, 355
  expansion, 33, 235, 334
  formed, 304-305
  longitudinal, 354
  open, 303-304
  warping, 355
Joists, 126, 131-132

Kerbs, 238, 355-356

Labour grades
  4, 23, 44, 61, 67, 75, 100, 130, 136, 139, 155, 176, 198, 212, 221, 233, 244, 274, 285, 292,

Labour grades (cont'd)
  312, 348, 365, 383, 394, 405,
  409, 424, 429, 436, 441, 450,
  468, 507, 518
Laminated beams, 107-108
Land drainage, 435-440
Landscaping, 421-443
Leadwork, 74, 85-87
Lean concrete, 350
Linings
  blockboard, 116-119
  chipboard, 116-119
  hardboard, 116-119
  insulation board, 117-119
  plywood, 117-118
Lintels
  precast concrete, 37
  steel, 137-138
Lighting, 545-546
Luminaires, 539-544

Macadam
  dense, 351
  dry-bound, 351
  open textured, 351
  wet-bound, 350
Manholes, 256-258, 336-340
Masonry, 374
Masonry paint, 230-231, 400-402
Measurement data, 577-581
Metalwork, 135-138
Mild steel
  fittings, 469-472, 477-482
  pipework, 469-472, 477-482
Mortar, 42, 43, 61, 362-363,
  364

Nails, 73-74, 98-100,
  380-381
Natural slates, 80-84
Natural stone walling, 64

Oil paint, 223-226, 396-398
Overflows, 182

Paints
  Artex, 227-228
  bituminous, 402-403
  coverage, 217-221, 394-395
  creosote, 229
  emulsion, 222-223, 398-400
  masonry, 230-231, 400-402
  oil, 223-226, 396-398
  polyurethane, 227-230, 402
  primers, 395-396
Palisade fencing, 242
Panel fencing, 240
Partitions, 101, 163-164, 174
Pavings
  brick, 236
  concrete, 237
  gravel, 236
Penstocks, 333
Piped supply systems, 447-465
Pipeline supports, 457-459
Pipework
  carbon steel, 473-475
  copper, 182-185, 451-455
  drainage, 251-255
  land drainage, 438-439
  mild steel, 469, 477-482
  polyethylene, 186
  rainwater, 177-178
Plant
  fuel consumption, 5-6
  grades, 4, 23, 68, 130, 233, 244.
    275, 292, 312, 348, 383, 405,
    424, 436
  outputs, 6, 274
Plasterwork, 165-173
Plywood
  casings, 117-118
  eaves, 103

Plywood (cont'd)
  fascias, 103
    gutters, 103
    linings, 117-118
    soffits, 103
    verges, 103
Polyethylene pipe, 186
Polyurethane, 227, 230, 402
Ponds, 442-443
Post and rail fencing, 239,
  410-411
Post and wire fencing, 240-241,
  411-412, 415-416
Power, 547
Precast concrete
  cappings, 37
  channels, 238, 356-357
  cills, 37
  copings, 37
  edgings, 237
  kerbs, 238, 353-356
  lintels, 37
Primers, 395-396
Protective layers, 407
Pumps, 443, 491-494

Quarry tiles, 146

Radiators, 501-502
Radio, 548
Rails, 114-116
Rainwater goods
  gutters, 179-180
  pipes, 177-178
Ready mixed concrete, 24-25
Reconstructed stone,
  slating, 85
  walling, 65
Refrigeration units, 502-506

Reinforcement
  bar, 22, 32, 290-291, 301-302,
    353
  fabric, 23, 32-33, 235, 291,
    302-303, 353
Roads, 347-357
Roofing
  felt, 75
  slates, 72-73
  tiles, 71-72, 76-77
Roof lights, 95
Rubber tiles, 148
Rubble walling
  coursed, 62
  irregular coursed, 62
  random, 62

Sand sub-base, 234-235
Sanitary fittings, 194-195
Sawn softwood
  generally, 386-387
  flat roofs, 101
  floors, 101
  kerbs, 102
  partitions, 101
  pitched roof, 101
  strutting, 102
Screeds, 139-140, 142-145
Screws, 388-389
Seeding, soiling and turfing,
  423-428
Seedlings, 430
Shelving, 115-116
Shrubs, 432-433
Sills, 37
Site clearance, 7, 269-270
Site-mixed concrete, 25-26
Skirtings, 114-115
  roofing, 72-73

Slates
  fibre-cement, 78-79
  natural, 80-84
  reconstructed, 85
  roofing, 72-73
Sleeves, 460
Soffits, 103, 104-106
Soft spots, 280
Soil cement, 349
Soil pipes, 181
Spikes, 388
Sprinklers, 462-465
Stairs, 118
Starters, 537
Steel
  angles, 128-129
  bars, 135-136, 137
  beams, 125-127
  channels, 126
  columns, 125-127
  joists, 126, 131-132
  lintels, 137-138
  tees, 127-128
Stonework, 61-65
Stone walling,
  ashlar, 374
  coursed, 62, 377
  irregular coursed, 67, 376
  natural, 64
  random rubble, 62, 375
  reconstructed, 65
Stopcocks, 191, 442, 460-461
Structural steelwork, 125-134
Sub-bases,
  granular, 234-235
  hardcore, 234-235
  sand, 234-235
Switchboards, 553-554
Switchgear, 555

Tanking, 68, 407

Tanks, 193, 495
Taping, 553-554
Taps, 195
Tees, 127-128
Telecommunications, 548
Television, 548
Terrazzo, 147, 150
Test points, 554
Thermal board, 159-162
Thermoplastic tiling, 148
Tiling
  carpet, 150
  ceramic, 147, 173
  cork, 149
  hardwood block, 150
  quarry, 146
  roofing, 71-77, 76-77
  rubber, 148
  terrazzo, 147
  thermoplastic, 148
  vinyl, 149
Timber, 379-392
Transformers, 529-531
Transplants, 430
Trees and shrubs, 429-434
Tree sizes, 430-432
Trench widths, 244, 311, 436
Trunking, 522-524
Trussed rafters, 107
Turfing, 427-428

Useful addresses, 583-589

Valves, 191, 332, 461, 483-484
  485-486, 487
Ventilation system, 507-511
Vinyl flooring
  sheeting, 151
  tiles, 149
  verges, 103, 104-106